Math Practice the Singapore Way

Grade 2

A LEADER IN MATH FOR 30 years

Marshall Cavendish Education

Published by Marshall Cavendish Education

Marshall Cavendish Corp.
800 Westchester Ave., Suite N-641
Rye Brook, NY 10573
Website: www.mceducation.us

Originally published as Maths Practice Papers © 2002 Times Media Private Limited, © 2003, 2010 Marshall Cavendish International (Singapore) Private Limited, © 2014 Marshall Cavendish Education Pte Ltd

First published 2012
Reprinted 2016, 2018, 2020 (twice)

Other Marshall Cavendish Offices:
Marshall Cavendish International (Asia) Private Limited, 1 New Industrial Road, Singapore 536196 • Marshall Cavendish International (Thailand) Co Ltd. 253 Asoke, 12th Flr, Sukhumvit 21 Road, Klongtoey Nua, Wattana, Bangkok 10110, Thailand • Marshall Cavendish (Malaysia) Sdn Bhd, Times Subang, Lot 46, Subang Hi-Tech Industrial Park, Batu Tiga, 40000 Shah Alam, Selangor Darul Ehsan, Malaysia.

ISBN 978-0-7614-8034-1

Printed in Singapore

Introduction

Mathematics the Singapore Way is a term coined to refer to the textbook series used in Singapore schools. Math the Singapore Way focuses on problem solving, given that it is based on a curriculum framework that has mathematical problem solving as its focus. Math the Singapore Way also focuses on thinking, given that the Singapore education system is driven by the *Thinking Schools, Learning Nation* philosophy. Math the Singapore Way is also based on learning theories that provide clear directions on how mathematics is learned and should be taught.

Singapore mathematics textbooks, initial teacher preparation, and the subsequent professional development for teachers are based on helping teachers understand what to teach in mathematics and how to teach it.

While these books are not part of the formal classroom program, they provide selected groups of students with the necessary consolidation of skills. Some students require more opportunities to consolidate their basic concepts and skills, and these books are written with that goal in mind. Such learning materials, if prepared consistently with the fundamentals of Mathematics the Singapore Way, help learners review basic concepts through the use of visuals. In some cases, it may be necessary for some learners also to have access to concrete materials. It should be noted that good practice is not a matter merely of random repetition. Learners must be helped through careful scaffolding. Good practice consists of careful variations in the tasks learners are given.

I hope this series of books is able to provide the necessary help for learners who need to be challenged beyond basic concepts and skills.

Yeap Ban Har
Marshall Cavendish Institute

Preface

Math Practice the Singapore Way is a series of five books with exercises that adhere closely to the latest Math syllabus in Singapore.

The topics in this book are carefully arranged to follow the sequence of the topics in the students' basic texts, thus enabling them to use the exercises as further practice to gain mastery of mathematical skills. The exercises can also be used to supplement teachers with a thorough and reliable program of testing that provides information for necessary re-teaching.

In mathematics, constant review of concepts is essential. Therefore, the built-in test yourself exercises will refresh students' memories so that no basic skills or concepts will be forgotten.

The variety of sample exam questions found in the final tests at the end of the book will help pupils face their final math tests with confidence.

Notes pages are provided at the back of the book for those who require more space to work out solutions to the problems.

Contents

UNIT 1 Notes

Concept: The place value of a digit refers to the value of the digit according to the "place" the number is in.

In 624,

- the digit 4 is in the ones place
- the digit 2 is in the tens place
- the digit 6 is in the hundreds place

Hundreds	Tens	Ones
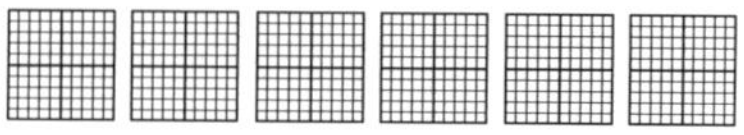	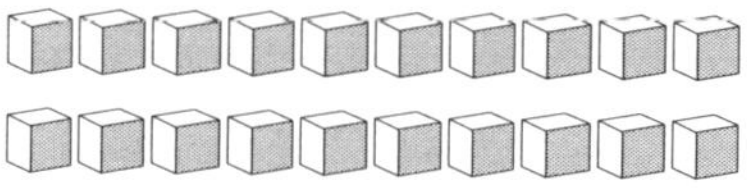	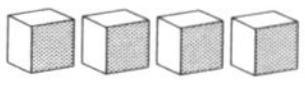
It stands for a value of 6 hundreds or 600	It stands for a value of 2 tens or 20	It stands for a value of 4 ones or 4

We can write 624 in words as:
Six hundred and twenty-four
624 = 600 + 20 + 4
624 = 6 hundreds 20 tens and 4 ones

Concept: We can count on by ones, tens, and hundreds.

i) Count On by Ones

+1 +1 +1 +1 +1

22	23	24	25	26	27	28	29	30

ii) Count On by Tens

+10 +10 +10 +10 +10

220	230	240	250	260	270	280	290	300

iii) Count On by Hundreds

+100+100+100+100+100

100	200	300	400	500	600	700	800	900

Concept: Comparing and ordering numbers.

We work from left to right when we compare or order numbers.

1.Which number is larger?

Hundreds	Tens	Ones
7	2	5
5	7	8

First, compare the digits in the hundreds place.
7 hundreds is **larger** than 5 hundreds.
So, 725 is **larger** than 578.

2. Arrange the numbers 695, 831, and 814 in order from smallest to **largest**.

Hundreds	Tens	Ones
6	9	5
8	3	1
8	1	4

Step 1
First, compare the digits in the hundreds place.
6 hundreds is the **smallest**, so 695 is the **smallest** number.

Step 2
Since 831 and 814 have the same digit in the hundreds place, we will compare their digits in the tens place.
1 hundred is smaller than 3 hundreds, so 831 is the **largest** number.

695 814 831

smallest

Concept: Number Patterns

We find out how numbers are related to one another in a pattern in order to find the missing numbers.

Complete the number patterns.
a. 53, 56, 59, ___, ___
b. 491, 484, 477, ___, ___

a.The difference between two numbers in the pattern is "+3."

+3 +3 +3 +3

53, 56, 59, 62, 65

b.The difference between two numbers in the pattern is "–7."

–7 –7 –7 –7

491, 484, 477, 470, 463

UNIT 1

Numbers to 1,000

POINTS
/50

1. Count the squares and write the number in the blank. (1 point)

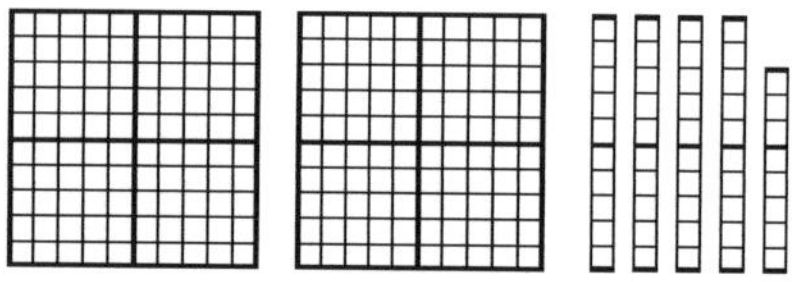

______

2. Circle the largest number. (1 point)

726 740 704 711

3. Circle the smallest number. (1 point)

384 512 290 400

4. Place an X (✗) by the one that is not equal to 472 in the box. (2 points)

(a) Four hundred and seventy-two ☐

(b) 400 + 70 + 2 ☐

(c) 400 + 20 + 7 ☐

5. Complete the following number sentence. (1 point)

685 = 600 + ______ + 5

6. Fill in the blanks. (4 points)

294 = ______ + 90 + 4

______ + 400 = 1,000

700 + 300 = ______

563 – ______ = 503

7. Write the answers to the questions in the blanks. (4 points)

What number is 10 more than 386? ________

What number is 10 less than 537? ________

What number is 100 more than 185? ________

What number is 100 less than 796? ________

8. Write the answers to the questions in the blanks. (4 points)

Which is larger, 563 or 653? ________

Which is larger, 1000 or 999? ________

Which is smaller, 909 or 910? ________

Which is smaller, 213 or 231? ________

9. Write the answers to the questions in the blanks. (3 points)

In 385, the numeral 3 stands for ________.

In 601, the numeral 1 stands for ________.

In 749, the numeral 4 stands for ________.

10. Match the numbers on the kites to the numbers on the students. (5 points)

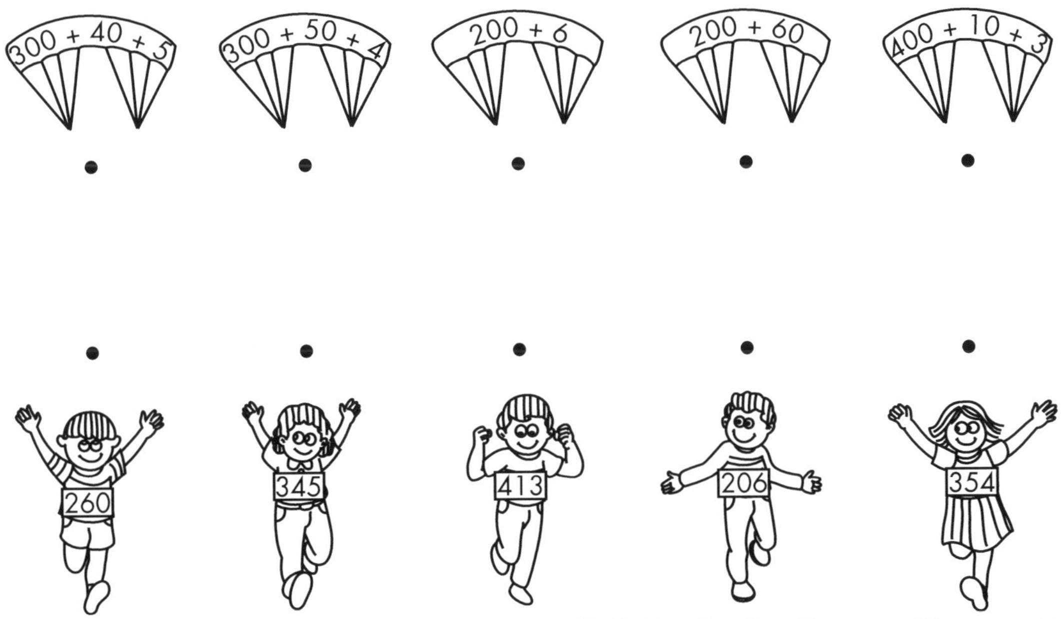

11. Write the following in numbers. (4 points)

Two hundred and twelve	
One hundred and six	

12. Write the numbers in words. (4 points)

813	
904	

13. Arrange these numbers in order beginning with the smallest.
Write your answers on the ladder steps.

276, 300, 259, 208

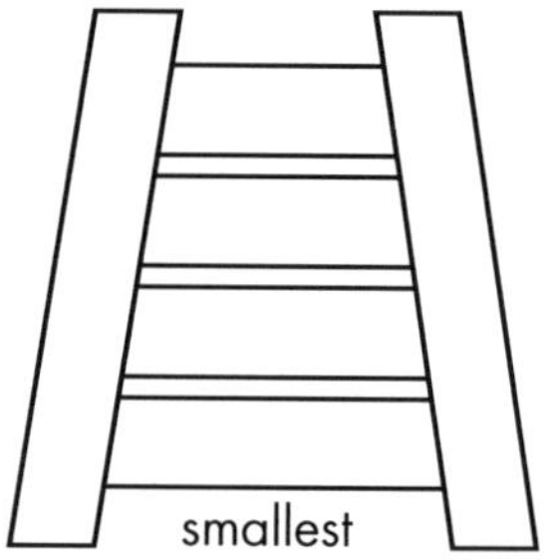

(2 points)

14. Arrange these numbers in order, beginning with the largest.
Write your answers in the balloons.

534, 576, 538, 580

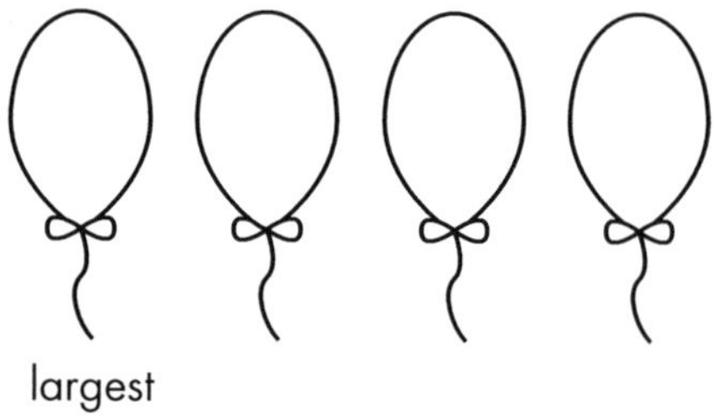

(2 points)

15. Write the answers in the blanks. (6 points)

90 and 4 make ________.

________ is 4 more than 90.

90 is 4 less than ________.

16. Write the answers in the blanks. (6 points)

100 and 8 make ________.

________ is 100 more than 8.

8 is 100 less than ________.

UNIT 2 Notes

Concept: Addition and subtraction of numbers up to 3 digits.

We add numbers together to find their total. This total is called the sum.

1. To add two 2-digit numbers, we add the ones first, followed by the tens. Find the sum of 43 and 38.

```
   T O
   ¹4 3
 + 3 8
   8 1  ——→ 8 + 3 = 11
                  = 1 ten and 1 one
            Regroup 10 ones as 1 ten
```

   The sum of 43 and 38 is 81.

2. To add two 3-digit numbers, we add the ones first, followed by the tens, and then the hundreds.
   Find the sum of 347 and 425.

```
   H  T  O      Step 1: Add the Ones
   3 ¹4  7      7 + 5 = 12
 + 4  2  5            = 1 ten and 2 ones
   7  7  2      Regroup 10 ones as 1 ten
                Step 2: Add the Tens
                1 + 4 + 2 = 7
                Step 3: Add the Hundreds
                3 + 4 = 7
```

   The sum of 347 and 425 is 772.

Concept: We subtract a number from a **larger** number to find the difference.

1. To subtract two 2-digit numbers, we subtract the ones first, followed by the tens. Find the difference between 48 and 36.

```
   T O
   4 8
 – 3 6
   1 2
```

   The difference between 48 and 36 is 12.

2. To subtract two 3-digit numbers, we subtract the ones first, followed by the tens, and then the hundreds.

   Find the difference between 446 and 339.

```
   H  T   O     Step 1: Subtract the Ones
   4 ³4̸ ¹6     Regroup 4 tens as 3 tens
 – 3  3   9     and 10 ones
   1  0   7     10 + 6 = 16
                16 – 9 = 7
                Step 2: Subtract the Tens
                3 - 3 = 0
                Step 3: Subtract the
                Hundreds 4 - 3 = 1
```

   The difference between 446 and 339 is 107.

UNIT 2

Addition and Subtraction within 1,000

POINTS ___ / 50

Complete the following. (24 points)

1. $804 - 303 = \square$

2. $674 - 254 = \square$

3. $800 - 500 = \square$

4. $999 - 500 = \square$

5. $\begin{array}{r} 273 \\ +\ 384 \\ \hline \\ \hline \end{array}$

6. $\begin{array}{r} 196 \\ +\ 378 \\ \hline \\ \hline \end{array}$

7. $\begin{array}{r} 614 \\ -\ 253 \\ \hline \\ \hline \end{array}$

8. $\begin{array}{r} 811 \\ -\ 375 \\ \hline \\ \hline \end{array}$

9. $\begin{array}{r} 286 \\ 94 \\ +\ 171 \\ \hline \\ \hline \end{array}$

10. $\begin{array}{r} 347 \\ 138 \\ +\ 256 \\ \hline \\ \hline \end{array}$

11. $\begin{array}{r} 603 \\ -\ 257 \\ \hline \\ \hline \end{array}$

12. $\begin{array}{r} 800 \\ -\ 214 \\ \hline \\ \hline \end{array}$

Look at the pictures and answer the questions.

13. How many more beads are there in Bag A than Bag B?

(2 points)

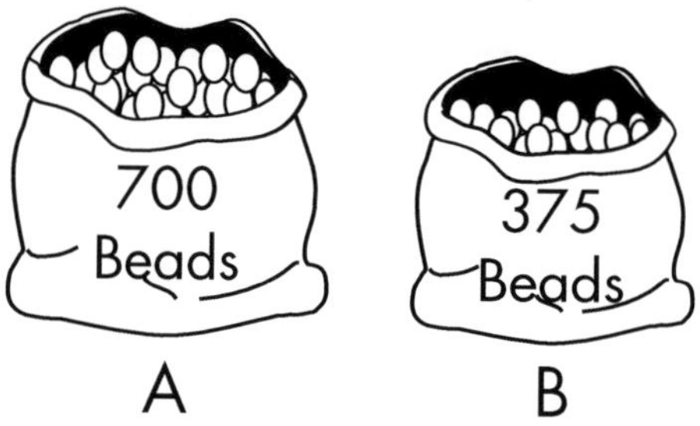

There are ________ more beads in Bag A than Bag B.

14. There are 942 students in a school.

685 of them are boys.

How many girls are there in the school? (3 points)

There are ________ girls.

15. There are 99 sheep in Pen A and 111 sheep in Pen B.

How many sheep are there in both pens? (3 points)

There are ________ sheep in both pens.

16. There are 86 red apples and 205 green apples in a box.

How many more green apples than red apples are there in the box? (4 points)

There are ________ more green apples than red apples.

17. There are 68 cows, 123 goats, and 39 horses in a field.
How many animals are there altogether? (4 points)

There are ________ animals altogether.

18. Maggie made 88 more footprints than Selena.
Selena made 188 footprints.
How many footprints did both of them make altogether?
(4 points)

19. Marcus sold 684 pieces of fruit last weekend.
236 were oranges and the rest were apples.
He sold 219 apples on Saturday.
How many apples did he sell on Sunday? (6 points)

UNIT 3

Notes

Concept: Building up the multiplication tables of 2, 3, 4, 5, and 10.

We can skip-count (count on) by 2 to help us remember the multiplication table of 2.

2, 4, 6, 8, 10, 12, 14, 16, 18, 20

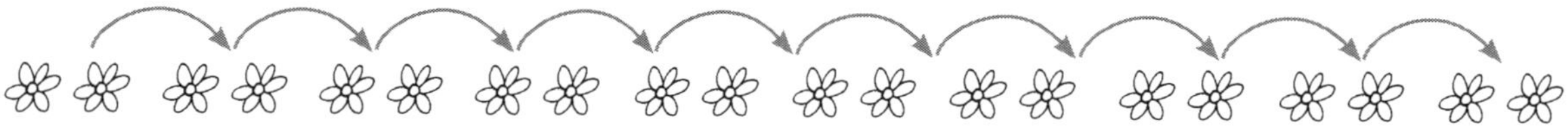

Concept: We can skip-count (count on) by 3 to help us remember the multiplication table of 3.

3, 6, 9, 12, 15, 18, 21, 24, 27, 30

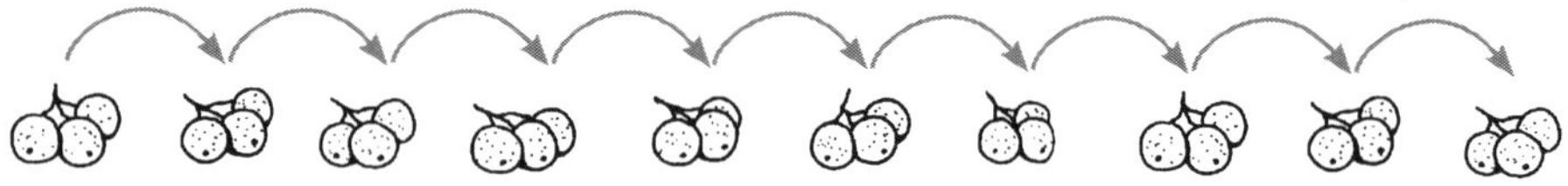

Concept: We can skip-count (count on) by 4 to help us remember the multiplication table of 4.

4, 8, 12, 16, 20, 24, 28, 32, 36, 40

Concept: We can skip-count (count on) by 5 to help us remember the multiplication table of 5.

5, 10, 15, 20, 25, 30, 35, 40, 45, 50

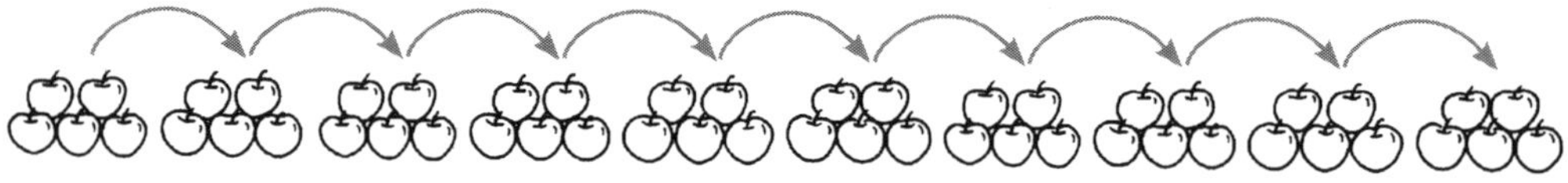

Concept: We can skip-count (count on) by 10 to help us remember the multiplication table of 10.

10, 20, 30, 40, 50, 60, 70, 80, 90, 100

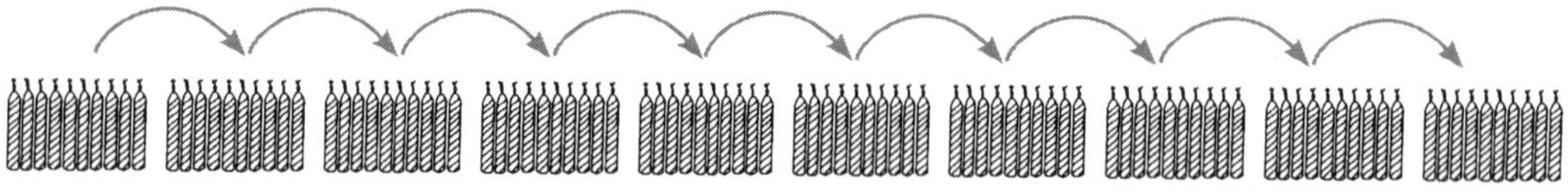

UNIT 3

Multiplication Tables of 2, 3, 4, 5, and 10

POINTS
/50

Complete the following problems.

1. 

7 threes = ☐

7 × 3 = ☐ (2 points)

2. 

4 groups of 5 = ☐

4 × 5 = ☐ (2 points)

3. Multiply 6 by 3.

3 times 6 = ☐

3 × 6 = ☐ (2 points)

4. Draw 2 sets of 3 oranges.

$2 \times 3 = \square$

There are ________ oranges altogether. (2 points)

5. Count in steps of 2.

2	4	6	8				

8 twos = $\square$

$8 \times 2 = \square$ (2 points)

6.

$\square \times \square = \square$

There are ________ candles altogether. (2 points)

7. There are 5 pastries on one plate.

How many pastries are on 4 plates?

$\square \times \square = \square$

There are ________ pastries on 4 plates. (2 points)

8. Complete the drawing to show 2 rows of 4 flowers.

2 rows of 4

Complete the drawing to show 4 rows of 2 flowers.

4 rows of 2

Now, write the amounts as number sentences.

$\square \times \square = \square$ $\square \times \square = \square$

$2 \times 4 = 4 \times \square$ (2 points)

9. Complete the following table. (3 points)

1 stool has 3 legs.

Number of stools	1	2	3	4	5	6	7
Number of legs	3						

10. (a)

3 rows of 5 = ☐

3 × 5 = ☐

(b)

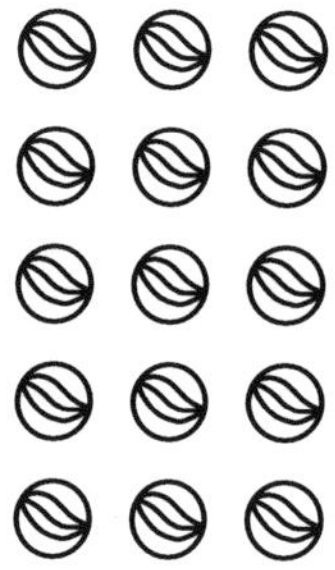

5 rows of 3 = ☐

5 × 3 = ☐ (2 points)

11. Complete the following. (4 points)

(a) $2 + 2 + 2 + 2 + 2 = \square \times 2$

Draw  (smiley faces) in the box to show how many there are altogether.

(b) $6 \times 3 = \square + \square + \square + \square + \square + \square$

Draw  (moons) in the box to show how many there are altogether.

12. Write two number sentences for the following pictures. (4 points)

(a) 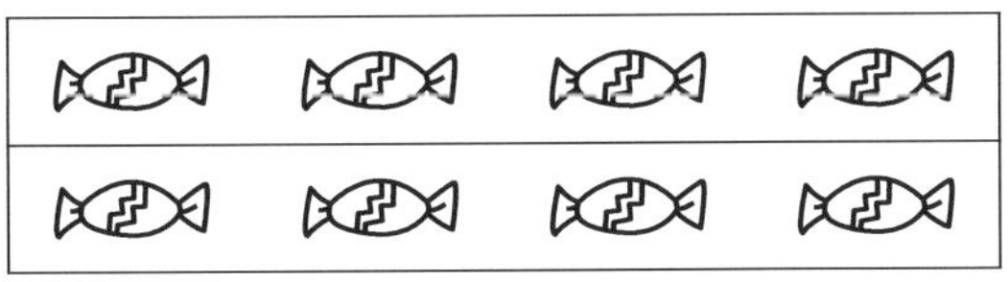

$\square \times \square = 8$

$\square \times \square = 8$

(b) 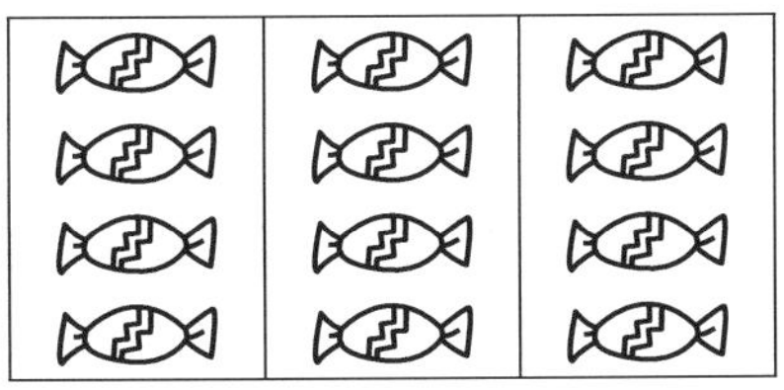

$\square \times \square = 12$

$\square \times \square = 12$

13. Add more checks (✓) to show 9 × 2 = 18. (1 point)

✓	✓							
✓	✓							

14. Add more checks (✓) to show 7 × 3 = 21. (1 point)

✓				
✓				
✓				
✓				
✓				
✓				
✓				

15. 1 airplane has 2 wings.
How many wings do 10 airplanes have?

(2 points)

☐ × ☐ = ☐

10 airplanes have ________ wings.

Read the questions carefully. Draw to solve each multiplication. Show all your work clearly.

16. Mr. Taylor eats 2 eggs every day.

    How many eggs does he eat in 7 days? (4 points)

    He eats ________ eggs in 7 days.

17. Elise has 8 vases.

    She puts 3 flowers into each vase.

    How many flowers does she have altogether? (4 points)

    She has ________ flowers altogether.

18. Betty had 6 shirts.

She sewed 4 pockets onto each shirt.

How many pockets did she have to sew onto 6 shirts? (4 points)

She had to sew ________ pockets on 6 shirts.

19. Ron and Bill bought 8 bags of carrots.

There were 3 carrots in each bag.

How many carrots did they buy altogether? (5 points)

They bought ________ carrots altogether.

UNIT 4

Notes

Concept: Recognizing the relationship between multiplication and division. Use of the division symbol (÷) to write a mathematical statement for a given situation. Multiplication and division within the multiplication tables.

We can find the total number of items arranged in equal groups by multiplication.

$2 + 2 + 2 = 6$

$3 \times 2 = 6$

$3 + 3 + 3 + 3 + 3 = 15$

$5 \times 3 = 15$

$4 + 4 = 8$

$2 \times 4 = 8$

$5 + 5 + 5 + 5 + 5 + 5 = 30$

$6 \times 5 = 30$

We can find the number of items shared equally among groups by division.

$8 \div 2 = 4$

$8 \div 2 = 4 \quad 2 \times 4 = 8$

$24 \div 4 = 6$

$24 \div 4 = 6 \quad 6 \times 4 = 24$

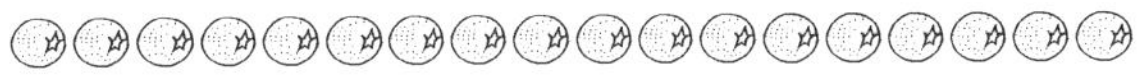

$18 \div 3 = 6$

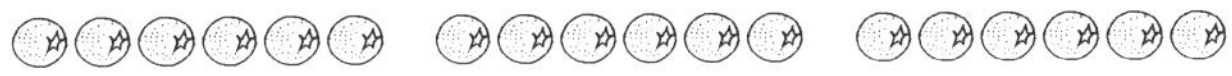

$18 \div 3 = 6 \quad 3 \times 6 = 18$

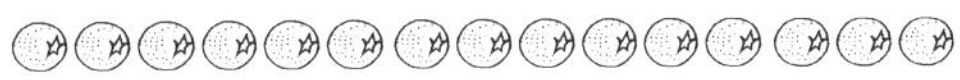

$15 \div 5 = 3$

$15 \div 5 = 3 \quad 3 \times 5 = 15$

$80 \div 10 = 8$

$80 \div 10 = 8 \quad 8 \times 10 = 80$

UNIT 4

Multiplication and Division

POINTS
/50

1. Draw lines to match the following problems with their answers. (4 points)

20 ÷ 2	9
24 ÷ 3	10
20 ÷ 4	8
90 ÷ 10	5

Fill in the boxes. (6 points)

2. (a) 6 × 3 = 3 × ☐ (b) 4 × 2 = ☐ × 4

3. (a) 2 × 10 = ☐ (b) 3 × 10 = ☐

4. (a) 9 × 3 = ☐ (b) 8 × 3 = ☐

5. Divide 20 toy cars into groups of 5. (1 point)

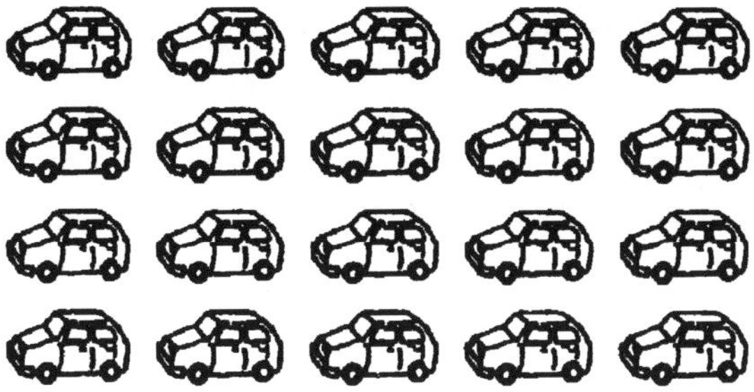

$20 \div 5 = \square$

There are ________ groups.

6. Use these pictures to write two multiplication and two division sentences.

(4 points)

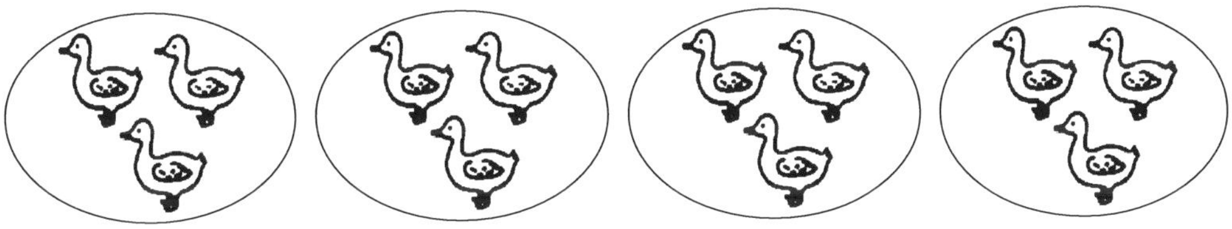

(a) $\square \times \square = \square$ $\quad$ $\square \times \square = \square$

(b) $\square \div \square = \square$ $\quad$ $\square \div \square = \square$

7. (a) $24 \div 3 = \square$ (1 point) (b) $20 \div 2 = \square$ (1 point)

8. There are 4 rice cakes in one box. (4 points)

(a)	Number of boxes	1	2		7	
(b)	Number of rice cakes	4		20		36

9. 

Put 15 bells into 3 equal groups.

☐ ÷ ☐ = ☐

There are ☐ bells in each group. (1 point)

10. 1 bicycle has 2 wheels. (3 points)

3 bicycles have ☐ wheels.

7 bicycles have ☐ wheels.

8 bicycles have ☐ wheels.

Fill in the boxes. (9 points)

11. 4 × 5 is 4 more than ☐.

12. 7 × 5 is 5 less than ☐.

13. 5 × 5 is 15 more than ☐.

14. (a) 3 × 4 = ☐

(b) 4 × 4 = ☐ + 4

(c) 7 × 4 = 20 + ☐

15. (a) $4 \times 8 =$ ☐

(b) $4 \times 9 = 32 +$ ☐

(c) $4 \times 7 = 32 -$ ☐

16. 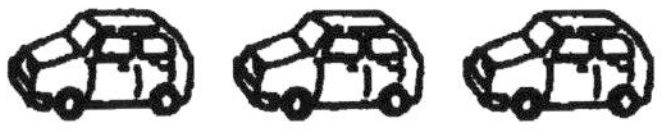 stand for 30 toy cars.

 stand for 30 + ☐ toy cars. (1 point)

Solve these problems.

17. Ali puts 3 birds in 1 cage.

How many cages does he need for 27 birds? (3 points)

☐ ÷ ☐ = ☐

He needs ________ cages.

18. Ms. Alonzo has 3 daughters.

She buys 4 new dresses for each daughter.

How many new dresses does she buy altogether? (3 points)

☐ × ☐ = ☐

She buys ________ new dresses in all.

19. Margie keeps 3 cages.

She has 2 rabbits in each cage.

How many rabbits does she have altogether? (2 points)

She has ____________ rabbits altogether.

20. Glenda used 3 spools of thread to make one dress for her doll.

She had 18 spools of thread.

She used all of the spools to make doll dresses.

How many doll dresses did she make? (2 points)

She made ____________ doll dresses.

21. Ms. Wallis bought 40 pencils for her class of 6 boys and 4 girls.
How many pencils did each student get? (2 points)

22. 80 students are divided into groups of 10.
There are 3 girls in each group.
(a) How many students are in each group?
(b) How many girls are there altogether? (3 points)

UNIT 5

Notes

Concept: We can use models to help us find the total.

i) when different groups of items are put together

34 + 54 = 88

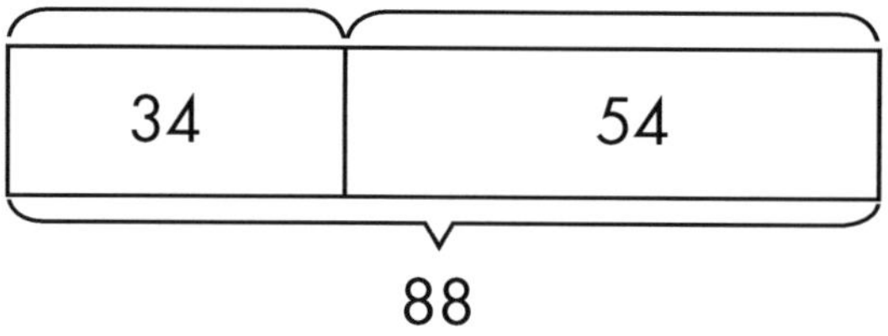

ii) when equal groups of items are put together

5 × 3 = 15

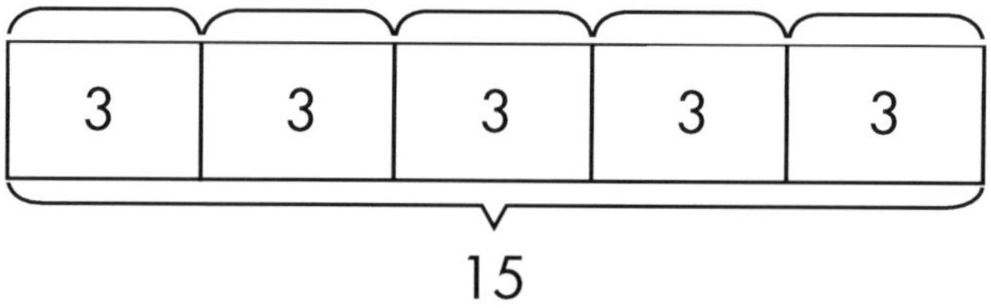

We can use models to help us find the difference when we subtract a smaller number from a larger number.

86 – 14 = 72

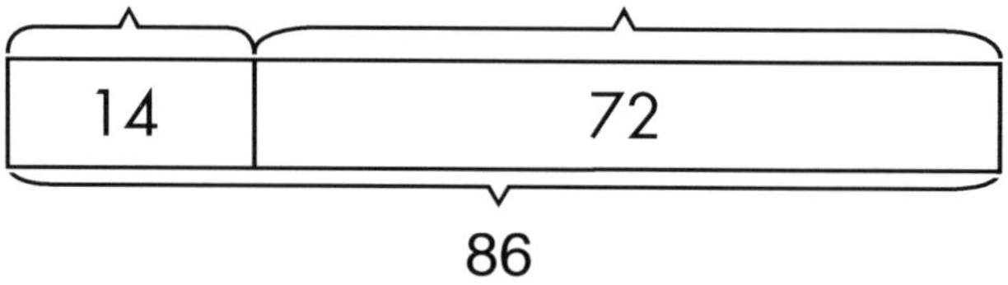

We can use models to find the number of items shared equally among groups by division.

27 ÷ 3 = 9

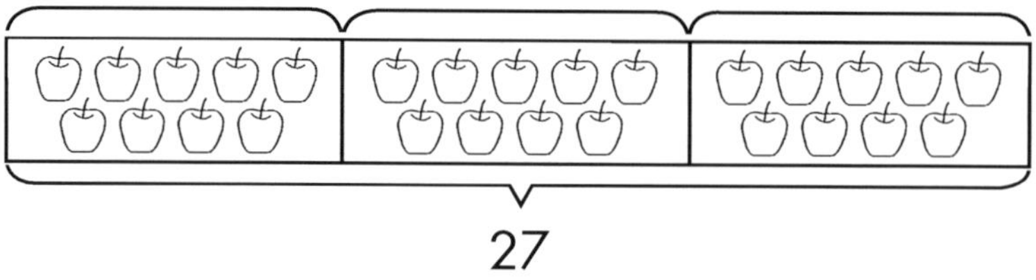

UNIT 5 Using Models

POINTS ___ / 50

Do the following sums using models.

1. Kim collects 20 seashells.

   Lollie collects 68 seashells.

   How many seashells do they collect altogether? (3 points)

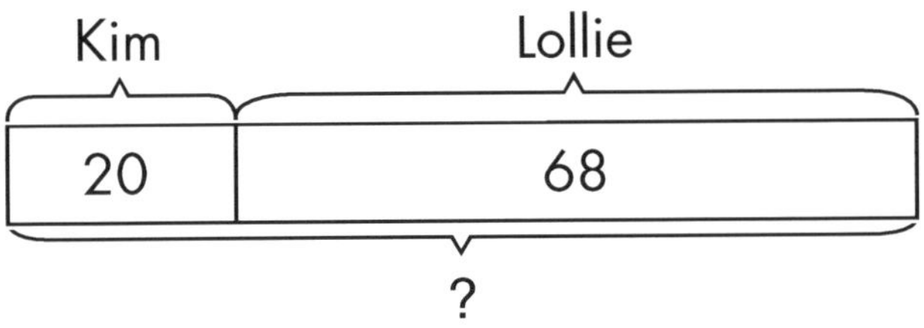

$20 + 68 = \square$

Kim and Lollie collect ☐ seashells altogether.

2. Nan baked 350 apple tarts.

   She sold 120 apple tarts.

   How many apple tarts did she have left? (3 points)

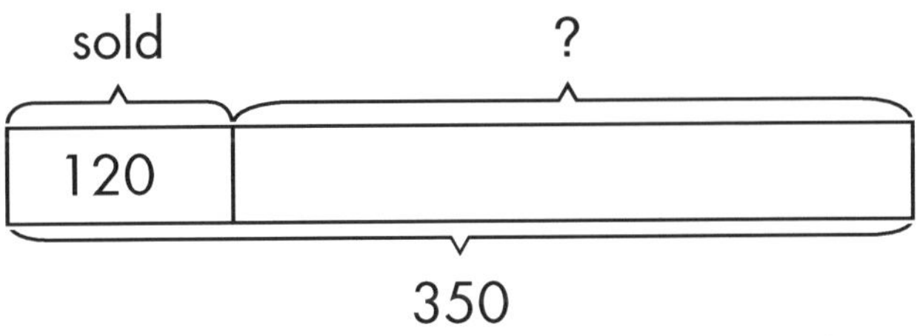

$350 - \square = \square$

She had ☐ apple tarts left.

3. There were 98 passengers on the bus.

   At the first stop, 65 passengers got off the bus.

   How many passengers were left on the bus? (3 points)

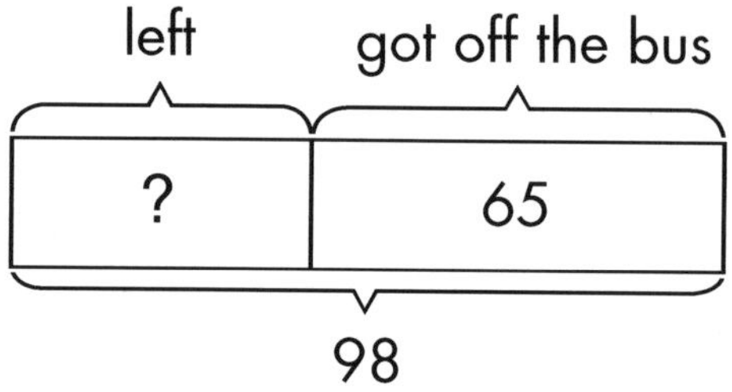

   _____ − _____ = _____

   There were _____ passengers left on the bus.

4. Jeff had 220 pears. He sold some pears and had 55 pears left.

   How many pears did he sell? (4 points)

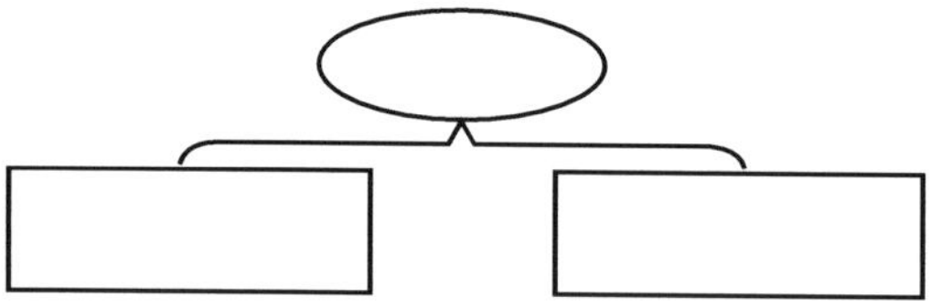

   Jeff sold ________ pears.

5. Jordan had 138 marbles. He won 83 marbles.

   How many marbles does he have now? (4 points)

   He has ________ marbles now.

6. Shawna had 420 stamps in her stamp album.
Her mother gave her 128 stamps on her birthday.
She received another 280 stamps from her friend.
How many stamps does Shawna have now? (4 points)

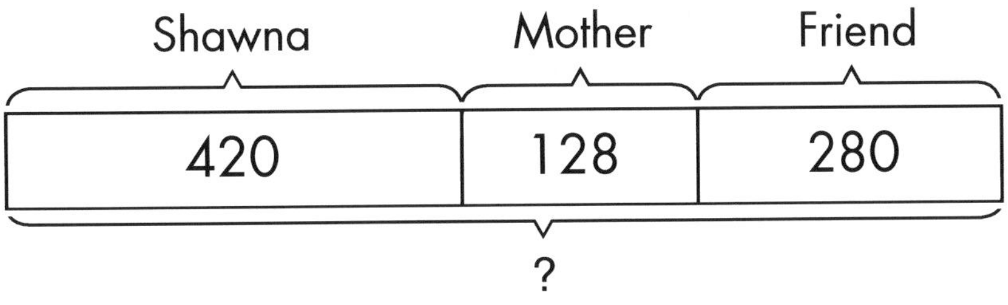

☐ + ☐ + ☐ = ☐

Shawna has ☐ stamps altogether.

7. Austin folded 96 paper planes on Monday.
He folded 69 paper planes on Tuesday.
He folded 66 paper planes on Wednesday.
How many paper planes did he fold in all? (4 points)

He folded ________ paper planes in all.

8. Diane sold 228 ice cream cones on Friday.
108 were chocolate, 74 were vanilla, and the rest were strawberry.

(a) How many chocolate and vanilla ice cream cones did she sell?

(b) How many strawberry ice cream cones did she sell?

(5 points)

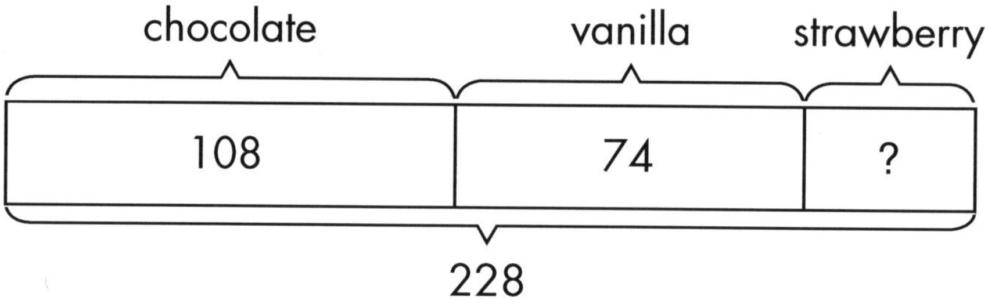

(a) ______ + ______ = ______

She sold ______ chocolate and vanilla ice cream cones.

(b) ______ − ______ = ______

She sold ______ strawberry ice cream cones.

9. Elaine painted 56 chairs.
She painted 17 more chairs than Daniel.
How many chairs did they paint altogether? (5 points)

They painted ________ chairs altogether.

10. Andrew had 49 fewer stickers than Joel.
Joel had 103 stickers.
How many stickers did they have in all? (5 points)

They had ________ stickers in all.

11. Jeff collected 78 popsicle sticks.
He collected 12 more popsicle sticks than Jen.
Jen collected 5 fewer popsicle sticks than Elise.
How many popsicle sticks did Elise collect? (5 points)

Elise collected ________ popsicle sticks.

12. Mr. Alvaro sold 350 tacos on Friday.
He sold 460 tacos on Saturday.
He sold 100 more tacos on Sunday than on Saturday.
How many tacos did he sell overall during the 3 days?
(5 points)

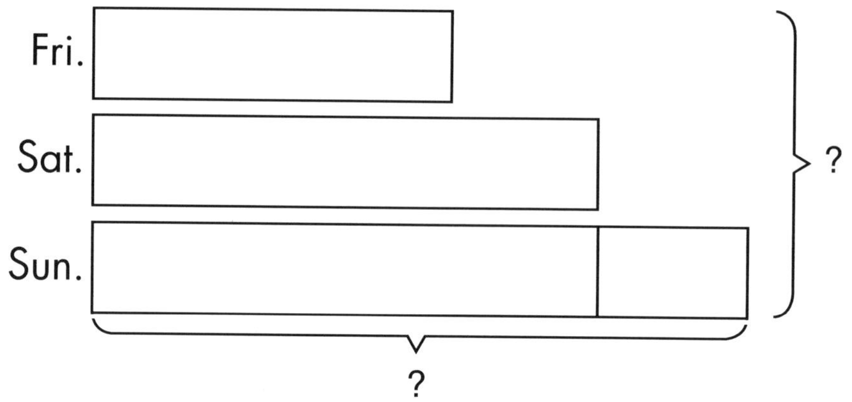

☐ + ☐ = ☐

He sold ☐ tacos on Sunday.

☐ ○ ☐ ○ ☐ ○ ☐

He sold ☐ tacos overall during the 3 days.

UNIT 6 Notes

Concept: Use of the appropriate measures and their abbreviations. We can measure length by meter (m), centimeter (cm), foot (ft.), and inch (in.). We usually use meter and foot for measuring long lengths, and inch and centimeter for short lengths.

We can estimate length.

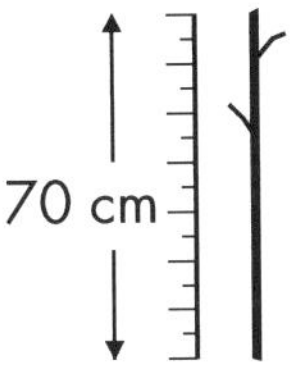

The stick is about 70 cm long.

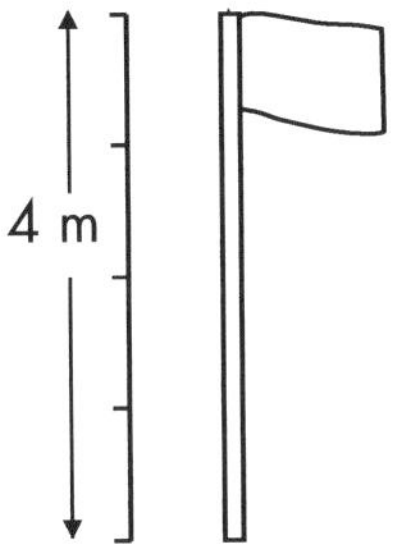

The pole is about 4 m long.

We can compare length.

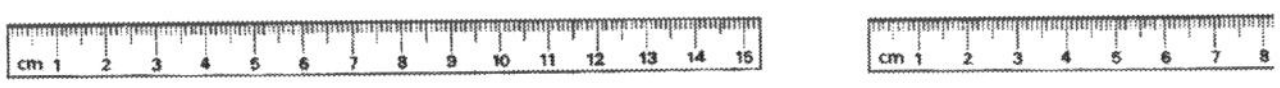

Ruler A is longer than Ruler B.

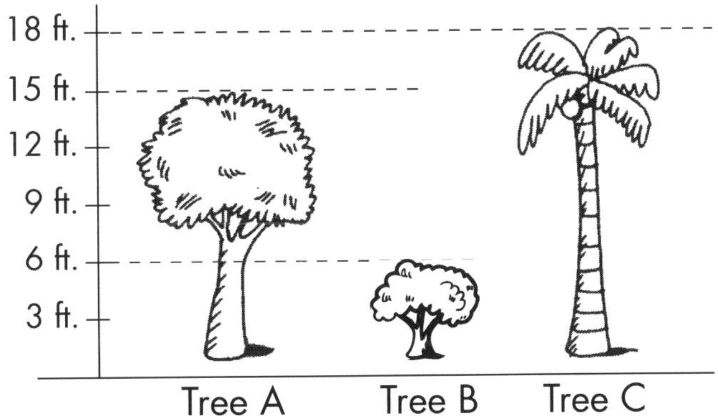

Tree C is the tallest.
Tree B is the shortest.
Tree A is taller than Tree B and it is shorter than Tree C.

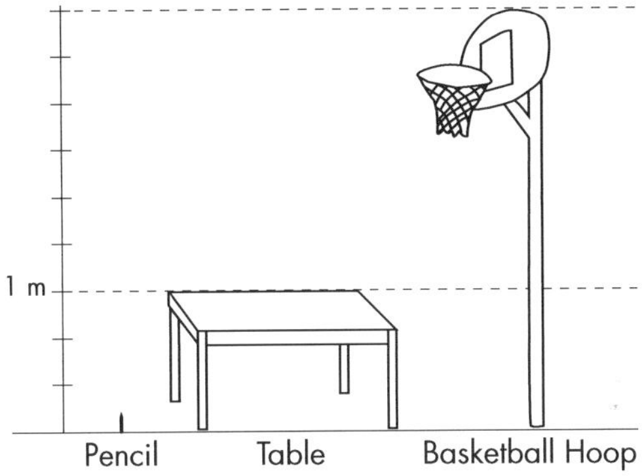

We can use a ruler to draw a straight line.

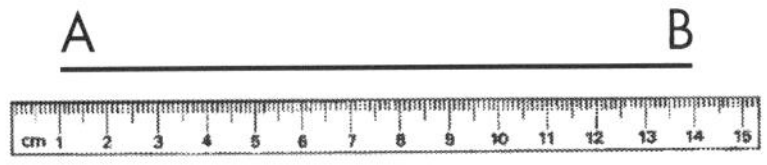

We can add, subtract, multiply, and divide measures of length in the same way we do for whole numbers.

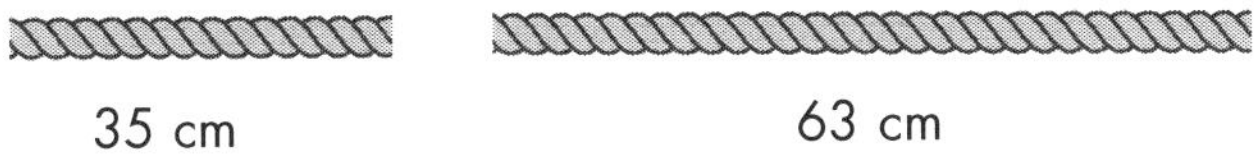

1. 35 cm + 63 cm = 98 cm

2. 18 – 13 = 5 m

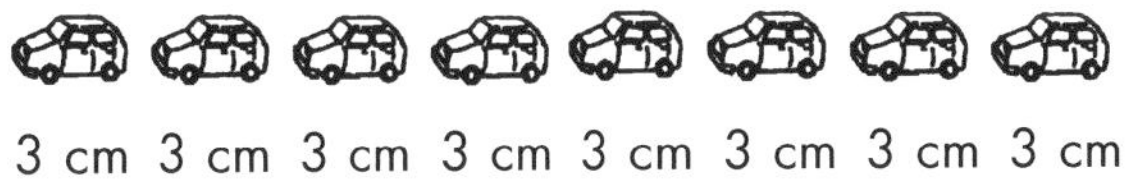

3. 8 × 3 = 24 cm

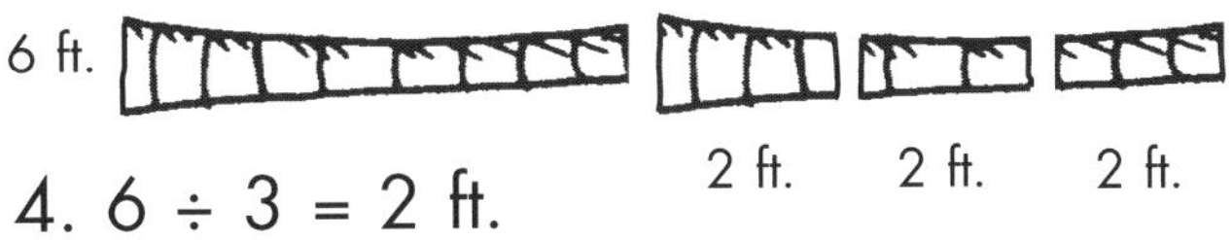

4. 6 ÷ 3 = 2 ft.

UNIT 6 Length

POINTS
/50

Put a check (✓) in the correct box. (4 points)

			Less than 1 yard	More than 1 yard
1.	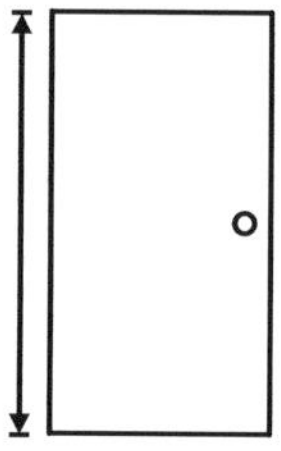	The height of my classroom door.		
2.		The width of my book.		
		The length of my book.		

3. This battery is about ________ centimeters long. (2 points)

Use a ruler to measure these items.

4. The crayon is about ________ inches long. (2 points)

5. The spoon is about ________ inches long. (2 points)

6. Fill in the blanks.

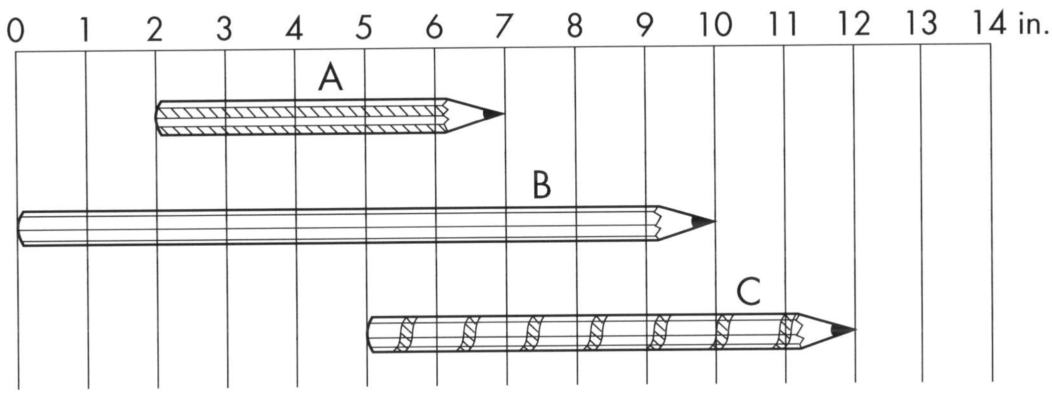

(a) Pencil A is about ________ inches long. (1 point)

(b) Pencil B is about ________ inches long. (1 point)

(c) Pencil C is about ________ inches long. (1 point)

(d) Pencil ________ is the longest. (1 point)

(e) Pencil ________ is the shortest. (1 point)

(f) The total length of the pencils is ________ inches. (1 point)

7. 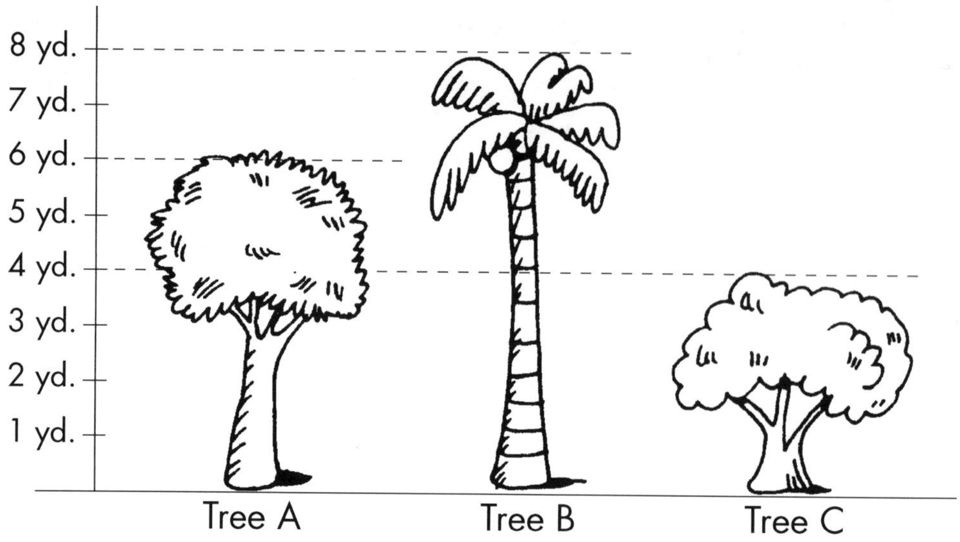

(a) Tree ________ is the tallest tree. (1 point)

(b) Tree ________ is the shortest tree. (1 point)

(c) Tree B is ________ yards taller than Tree A. (1 point)

(d) Tree C is ________ yards shorter than Tree A. (1 point)

(e) Tree ________ is 4 yards taller than Tree C. (1 point)

(f) The total height of the trees is ________ yards. (1 point)

8. 

(a) David is ________ centimeters tall. (1 point)

(b) Philip is ________ centimeters tall. (1 point)

(c) Winnie is ________ centimeters tall. (1 point)

(d) Winnie is ________ centimeters taller than Philip. (1 point)

(e) David is ________ centimeters shorter than Winnie. (1 point)

(f) The difference in height between the tallest and the shortest person is ________ centimeters. (1 point)

(g) Arrange the people from the shortest to the tallest.

________________________________ (1 point)

(h) The total height of the people is ________ centimeters. (1 point)

9. Vicky had a 50-inch ribbon.
She used 15 inches to tie up a present.
What was the length of ribbon left over? (3 points)

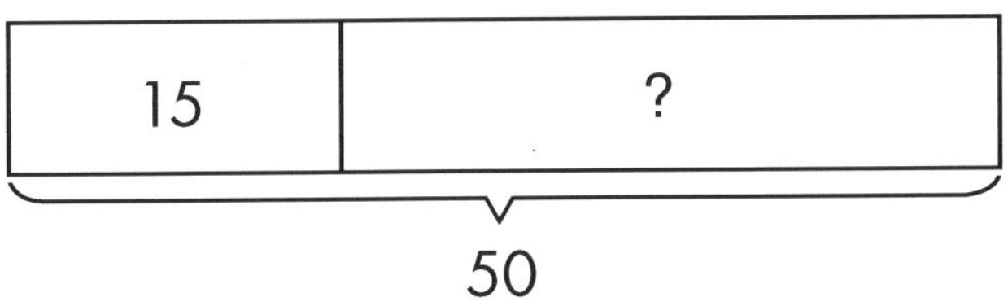

☐ – ☐ = ☐

☐ inches of ribbon were left over.

10. Lucy walked 400 yards to the post office and another 250 yards to the library.

What was the total distance she walked? (4 points)

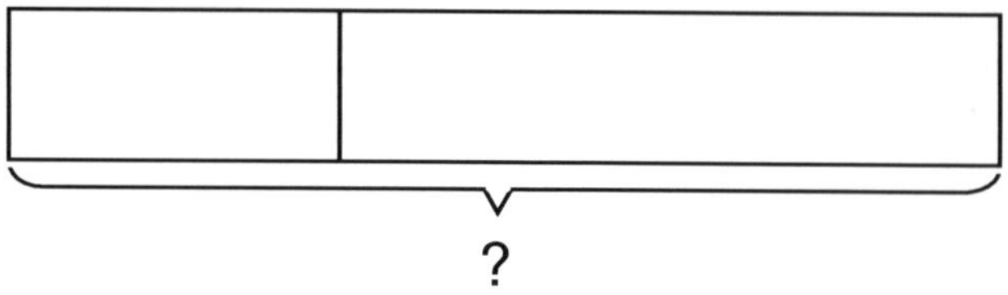

____ + ____ = ____

The total distance she had walked was ____ yards.

11. Mr. Brown is 12 centimeters taller than Ms. Brown is.

Ms. Brown is 4 centimeters shorter than her son.

Find (a) the height of Ms. Brown.

(b) the height of their son.

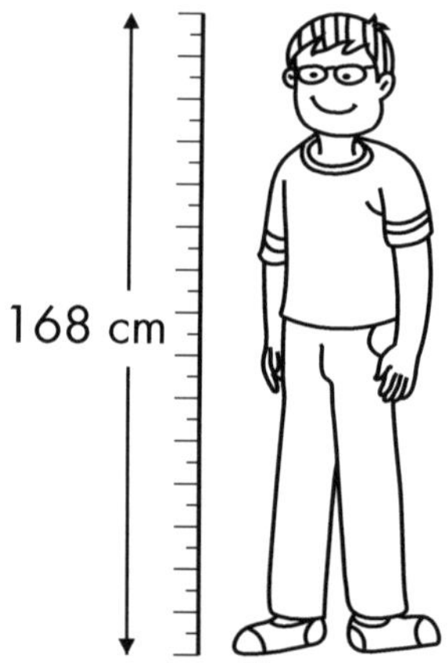

(4 points)

12. Calvin's kite and Cory's kite are flying in the sky.

The total height of the kites is 32 meters.

How high is Cory's kite flying?

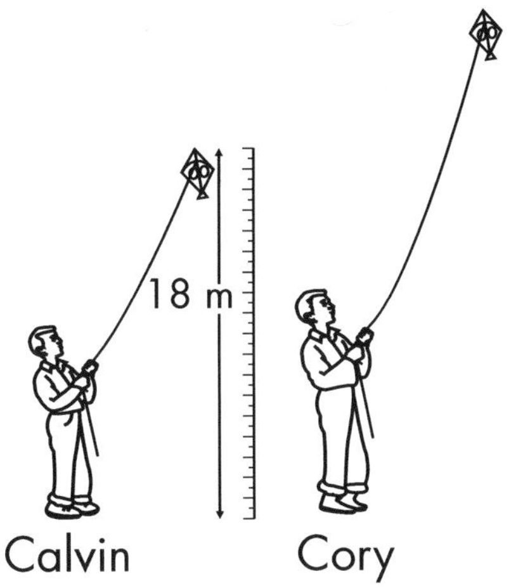

(4 points)

13. A piece of wire was cut into 2 pieces.

The first piece was 22 feet long.

The second piece was 5 feet shorter.

How long was the piece of wire at first? (5 points)

☐ − ☐ = ☐

The second piece of wire was ☐ feet long.

☐ + ☐ = ☐

The piece of wire was ☐ feet long at first.

UNIT 7 Notes

Concept: Use of the appropriate measures and their abbreviations. We can measure weight in kilograms (kg), grams (g), pounds (lb.), and ounces (oz.). We usually use kilograms and pounds for heavier items, and grams and ounces for lighter items.

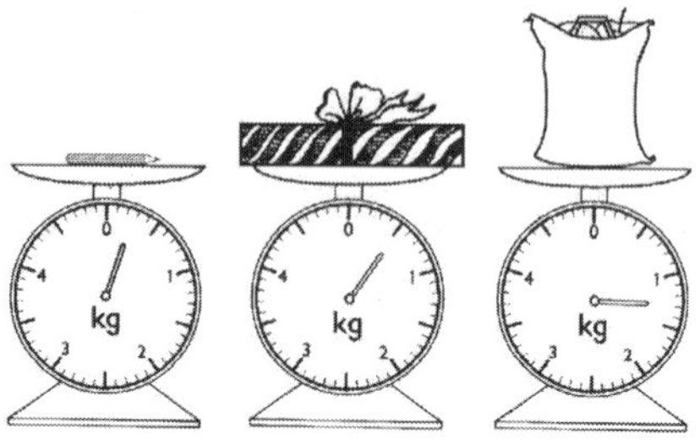

We can estimate weight.

1. The weight of the bag is about 900 g.

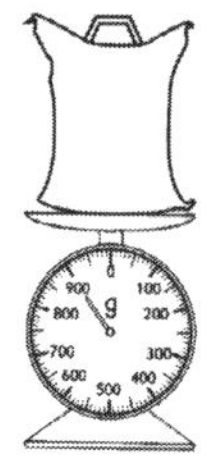

2. The weight of the bunch of bananas is about 2 kg.

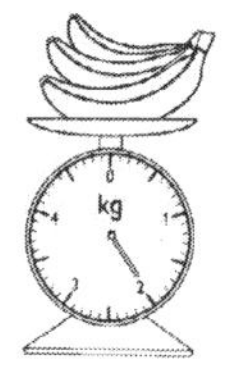

We can compare weight.

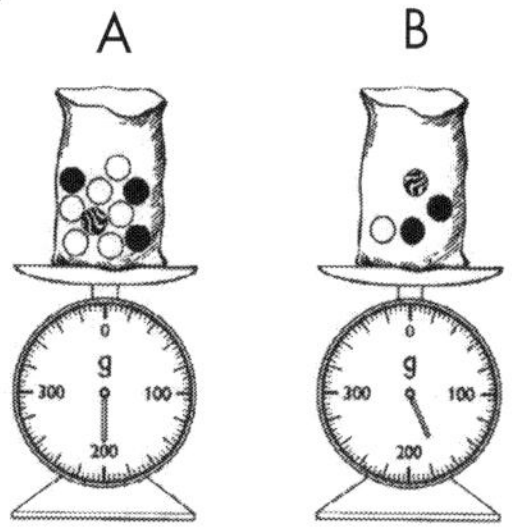

Bag A is heavier than Bag B.
Bag B is lighter than Bag A.

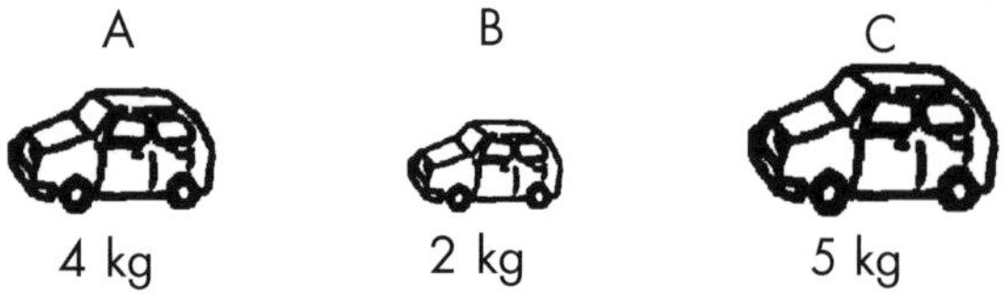

Toy C is the heaviest. Toy B is the lightest. Toy A is heavier than Toy B and it is lighter than Toy C.

We can add, subtract, multiply, and divide measures of weight in the same way we do for whole numbers.

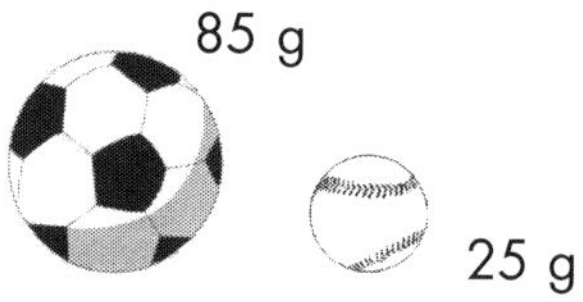

1. 85 g + 25 g = 110 g

2. 12 – 7 = 5 kg

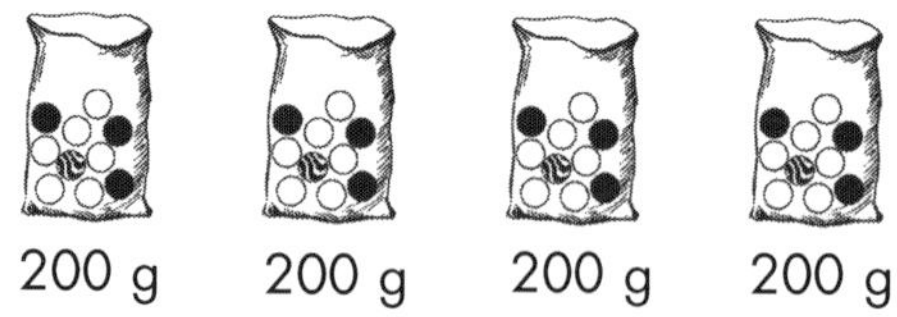

3. 4 × 200 = 800 g

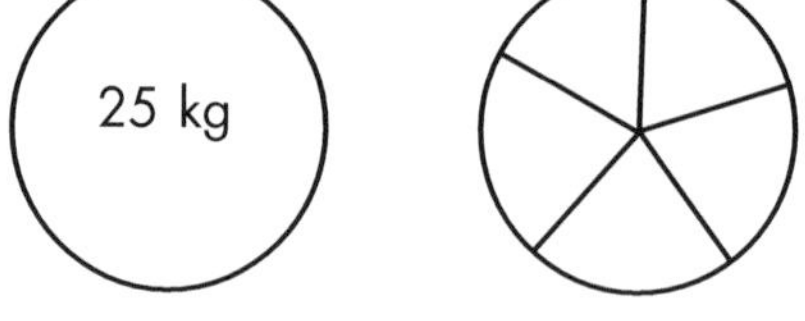

4. 25 ÷ 5 = 25 kg

UNIT 7

Weight

POINTS
/50

1. Write **lighter** or **heavier** in each blank.

   100 grams of nails are ________ than 1 kilogram of paper. (2 points)

   My bag full of books is ________ than my empty bag. (2 points)

2. A watermelon has a weight of 3 pounds.

   A cantaloupe has a weight of 2 pounds.

   The ________ is heavier than the ________. (1 point)

3. Put an X (✗) if the statement is **wrong**. (2 points)

   My teacher's heavy wooden chair has a weight less than 10 ounces. ☐

   Two girls of the same height must have the same weight. ☐

Fill in the blanks with the correct answers.

4. 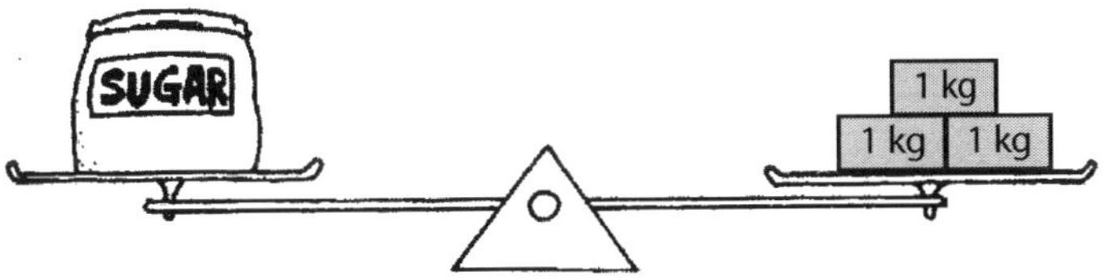

   The bag of sugar has a weight of about ________ kilograms. (2 points)

5.  The fish has a weight of about _________ pounds. (2 points)

Study the pictures. Then fill in the blanks.

David

Janet

6. David has a weight of about _________ kilograms. (2 points)

7. Janet has a weight of about _________ kilograms. (2 points)

8. Who is lighter, David or Janet? _________ (2 points)

9. They have a weight of _________ kilograms altogether. (2 points)

10. 

The chicken has a weight of about

________ grams. (2 points)

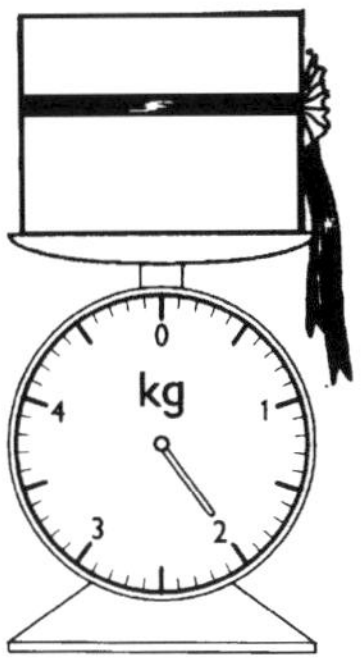

A

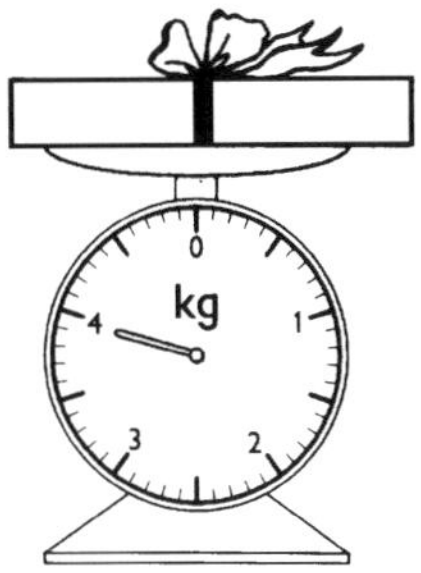

B

11. Package A has a weight of about ________ kilograms. (2 points)

12. Package B has a weight of about ________ kilograms. (2 points)

13. The weight of Package ________ is more than the weight of Package ________ by ________ kilograms. (2 points)

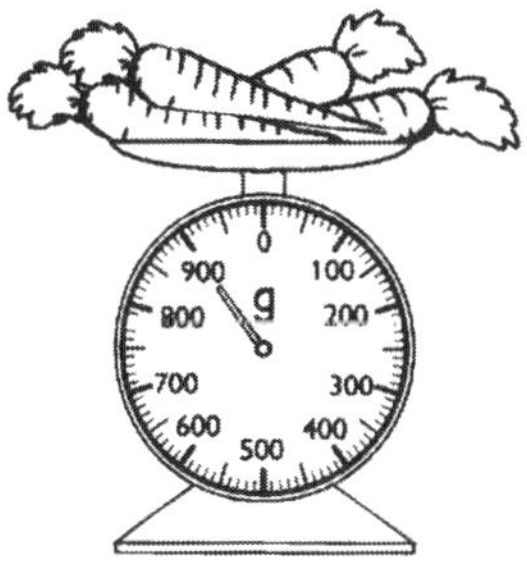

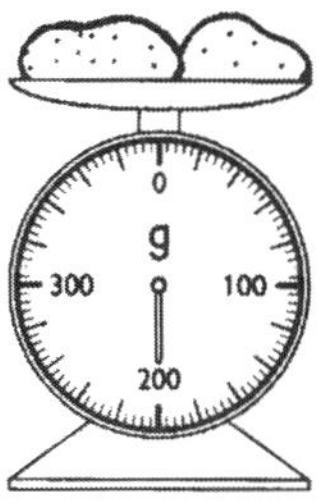

14. The carrots have a weight of about ________ grams. (2 points)

15. The squash have a weight of about ________ grams. (2 points)

16. The potatoes have a weight of about ________ grams. (2 points)

17. The squash is ________ grams heavier than the potatoes. (2 points)

18. Alvin has a weight of 37 pounds.

He is 15 pounds heavier than his younger sister.

What is the weight of Alvin's younger sister? (3 points)

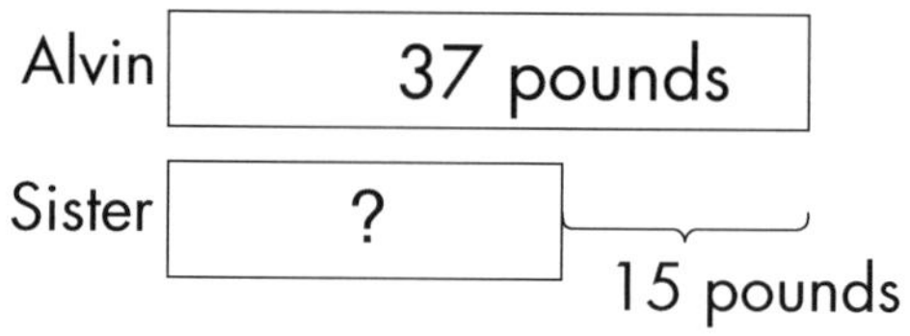

☐ ◯ ☐ = ☐

The weight of his sister is ________ pounds.

19.

James weighs the same as the total weight of the sacks of rice, potatoes, and onions.

Find James's weight. (4 points)

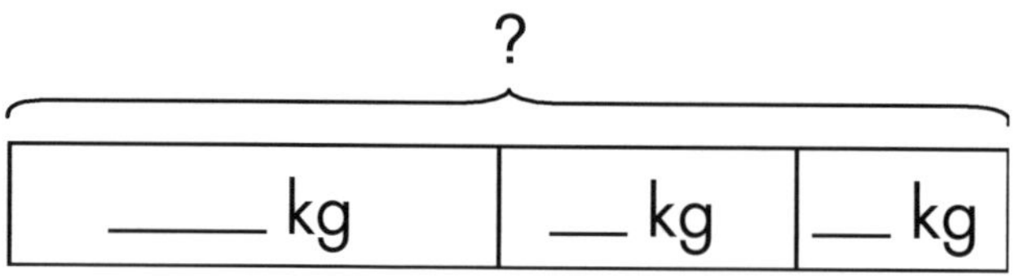

☐ ◯ ☐ ◯ ☐ = ☐

20. A watermelon is 15 ounces heavier than a honeydew.
The honeydew is 2 ounces lighter than a cantaloupe.
The watermelon has a weight of 4 pounds.
Find the weight of each fruit. (4 points)

21. Complete the problem with model drawings. (4 points)
A lamp has a weight of 328 ounces.
A table has a weight of 32 ounces more than the lamp.
(a) Find the weight of the table.
(b) Find the total weight of the table and the lamp.

(a)

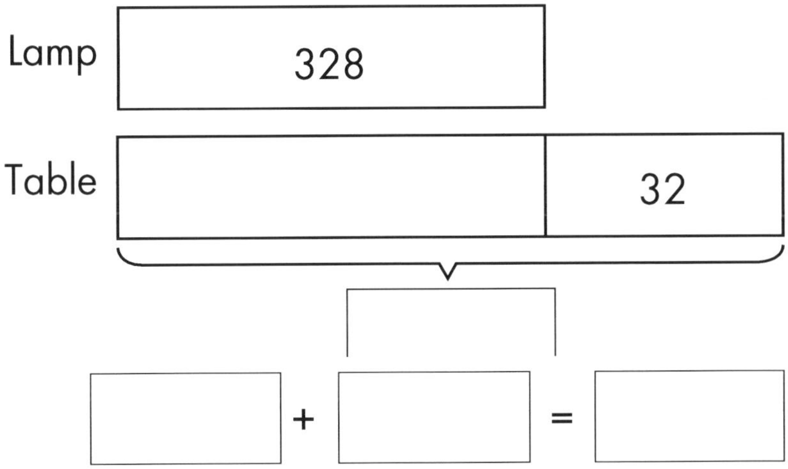

The table has a weight of ______ ounces.

(b)

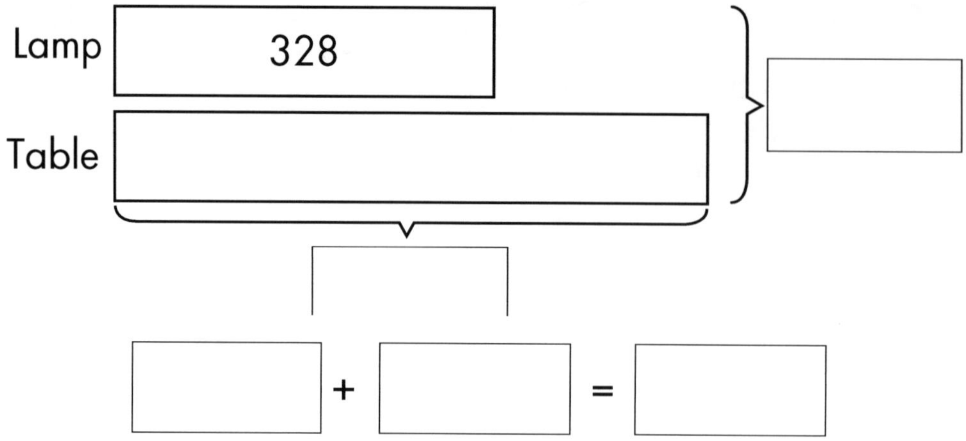

The total weight of the table and the lamp is ______ ounces.

UNIT

8 Notes

Concept: Use of the appropriate measures and their abbreviations.

We can measure volume in liters (L).

We can estimate volume. The volume of water in the tank is about 9 L.

We can compare volumes.

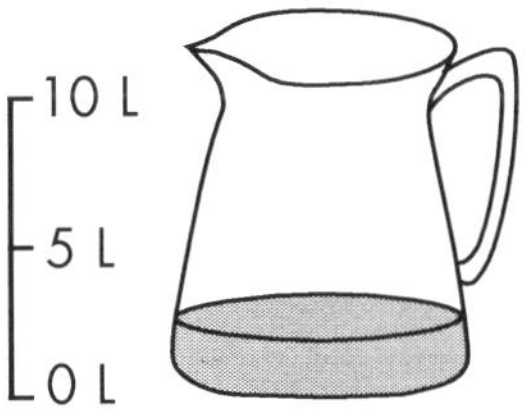

A

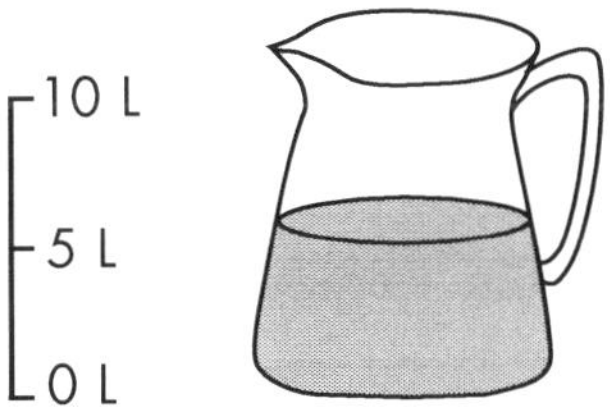

B

The volume of water in Jug B is more than the volume of water in Jug A.

Tank B has the largest volume of water in it.

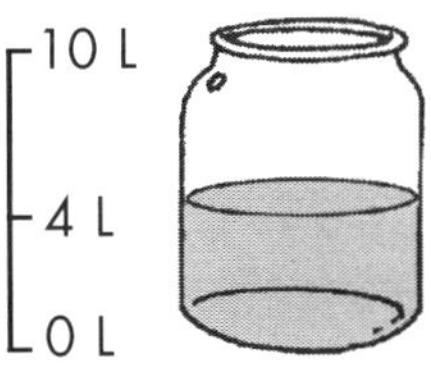

A

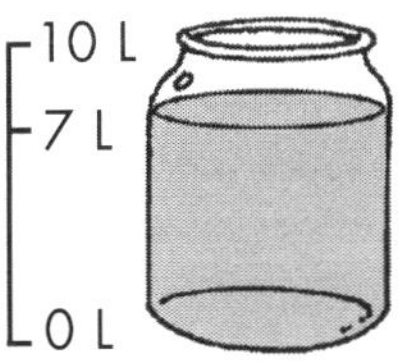

B

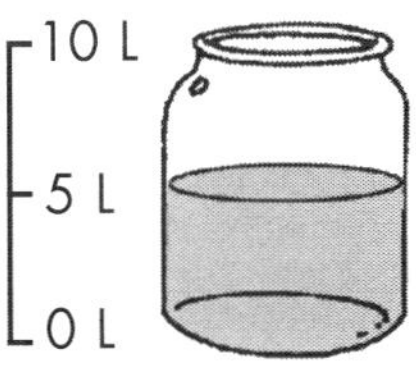

C

Tank A has the smallest volume of water in it.
The volume of water in Tank C is larger than the volume of water in Tank A and smaller than the volume of water in Tank B.

We can add, subtract, multiply, and divide measures of volume in the same way we do for whole numbers.

1. 7 L + 6 L = 13 L

2. 13 – 8 = 5 L

3. 5 × 2 = 10 L

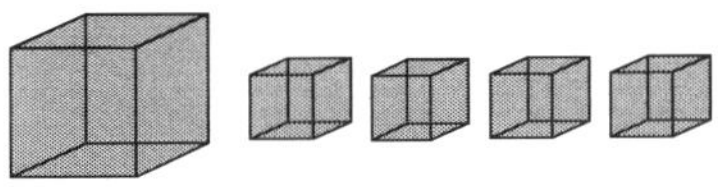

4. 16 ÷ 4 = 4 L

UNIT 8

Volume

POINTS: ___ / 50

1. Circle the glass that has more water. (1 point)

2. Circle the container that has less water. (1 point)

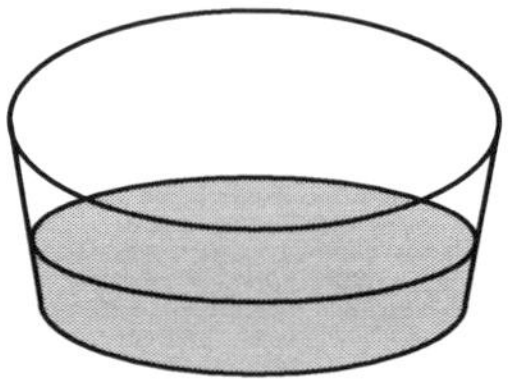 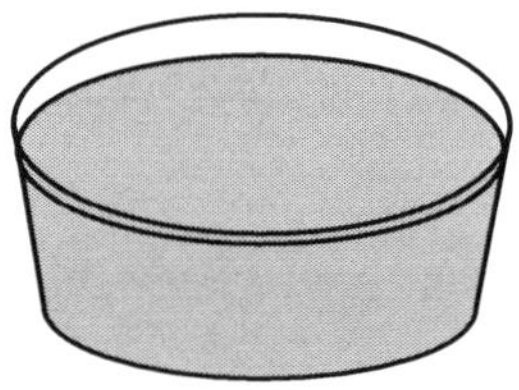

Study the pictures. Then, fill in the blanks. (1 point)

3. Which container has the largest volume of water? (2 points)

____

4. Which container has the smallest volume of water? (2 points)

____

Three bottles were full of water.

The water in each bottle was emptied into each glass.

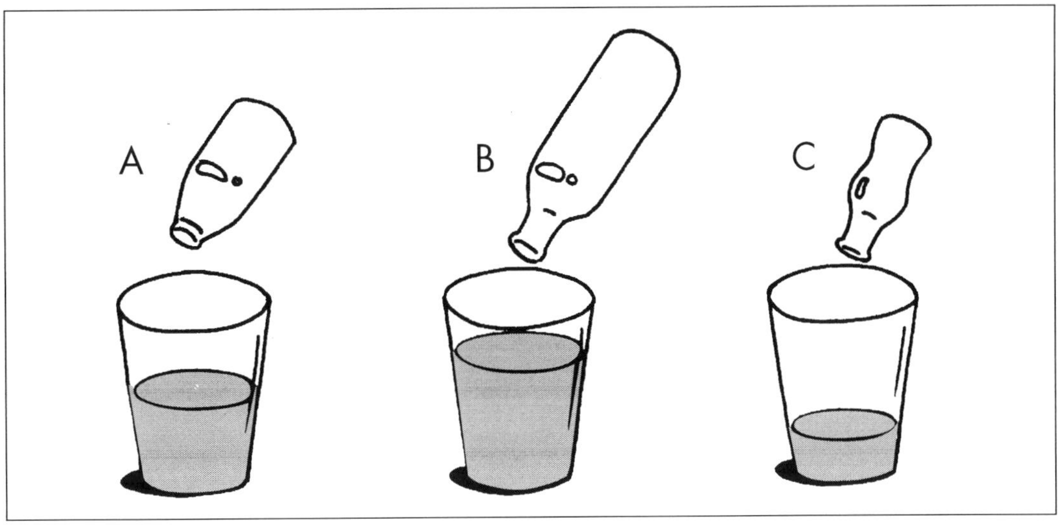

5. Bottle ________ has the smallest capacity. (2 points)

6. Bottle ________ has the largest capacity. (2 points)

7. Fill in the blank.

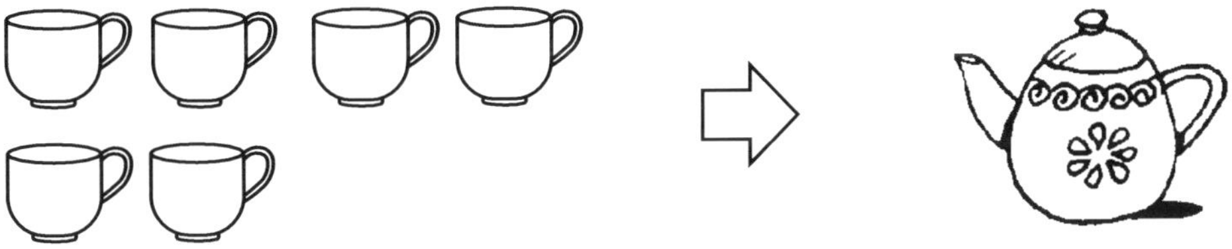

The volume of tea in the teapot is the same as ________ cups of tea. (2 points)

8. Study the pictures.
Then, fill in the blanks.
The glasses are emptied into each container.

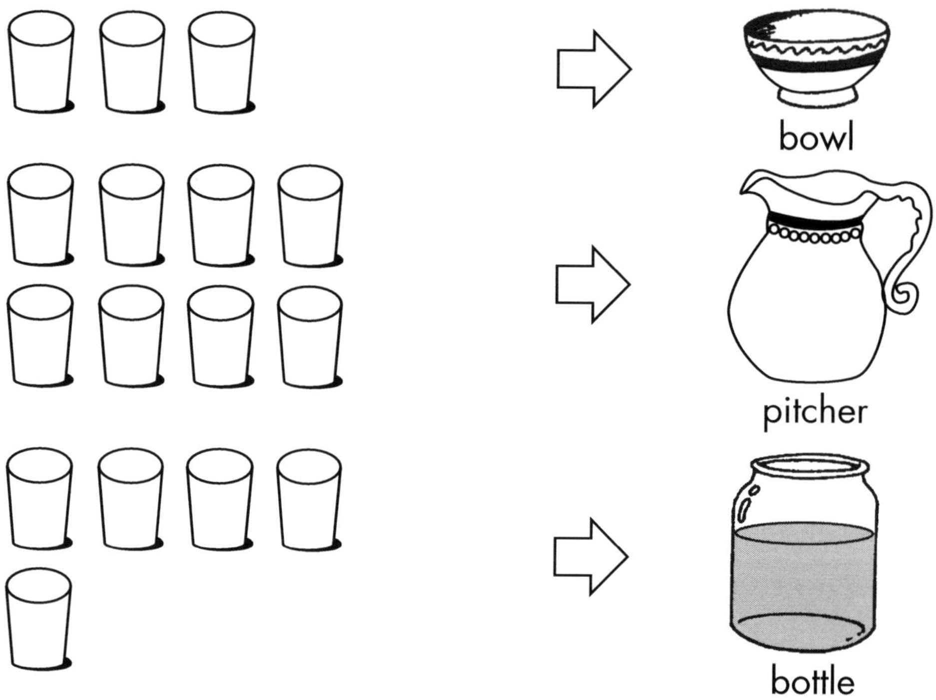

(a) The volume of water in the bowl is the same as
_________ glasses. (2 points)

(b) The volume of water in the pitcher is the same as
_________ glasses. (2 points)

(c) The volume of water in the bottle is the same as
_________ glasses. (2 points)

(d) The _________ has the largest volume of water.
(2 points)

9. The picture below shows a drum containing 50 liters of water, an empty pail, and an empty bottle.

(a) The volume of water in the drum is the same as how many pails full of water? ________ (2 points)

(b) The volume of water in the drum is the same as how many bottles full of water? ________ (2 points)

(c) If you fill the pail and the bottle with water from the drum, how many liters have been poured out of the drum? ________ (2 points)

(d) How many more liters of water are in the pail than in the bottle? ________ (2 point)

(e) How many liters are left in the drum? ________ (2 points)

Solve these problems.

10. Ms. Pierce has two tanks of water.

    One tank has 14 quarts of water in it.

    Another one has 17 quarts of water in it.

    Find the total amount of water in both tanks. (3 points)

    The total amount of water in both tanks is ________ quarts.

11. A fish tank is filled with 2 big pails of water and 1 small pail of water.

    A big pail can hold 20 quarts of water.

    A small pail can hold 14 quarts of water.

    How much water can the fish tank hold? (4 points)

12. Oil drum A can hold 190 gallons of oil.

It holds 70 gallons more oil more than drum B.

(a) How much oil can drum B hold?

(b) How much oil can both drums hold? (4 points)

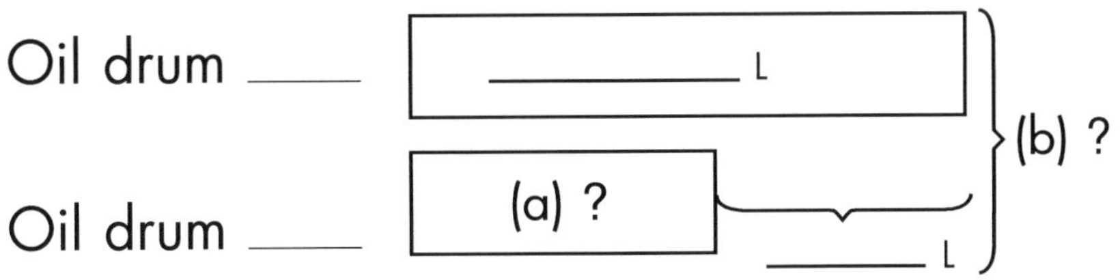

(a) ☐ ○ ☐ = ☐

Oil drum B can hold ________ gallons of oil.

(b) ☐ ○ ☐ = ☐

Both drums can hold ________ gallons of oil.

13.

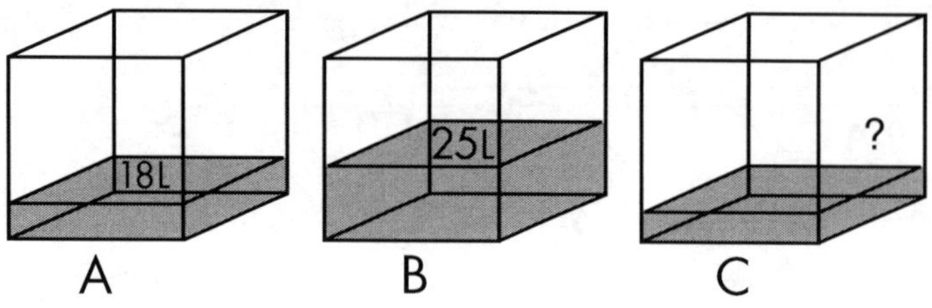

There are three tanks of water.

Tank B has 8 more liters of water than tank C.

How much water is in the three tanks altogether? (4 points)

14. A tank contained 188 gallons of water.

Lacie poured some water out of the tank into another container. 128 gallons of water were left in the tank.

(a) How many gallons of water were poured into another container?

(b) She poured the amount of water into two pails. One pail had 26 gallons of water. How much water was in the other pail? (4 points)

UNIT 9

Notes

Concept: Addition and subtraction involving 3-digit numbers and ones.

1. We can add ones to a 3-digit number mentally by adding a 10 and then subtracting the extra ones.

Add 324 and 7.
324 + 10 = 334
334 - 3 = 331

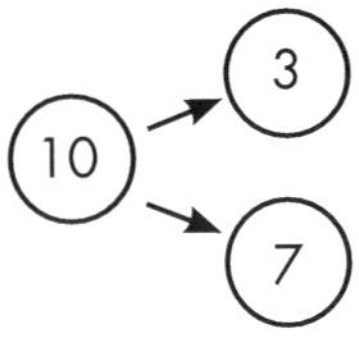

2. We can subtract ones from a 3-digit number mentally by subtracting 10 and then adding the extra ones back.

Subtract 6 from 562.
562 - 10 = 552
552 + 4 = 556

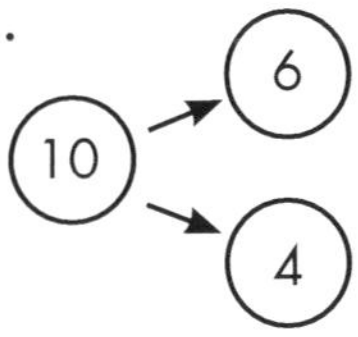

Concept:
Addition and subtraction involving 3-digit numbers and tens.

1. We can add tens to a 3-digit number mentally by adding 100 and then subtracting the extra tens.

Add 358 and 70.
358 + 100 = 458
458 - 30 = 428

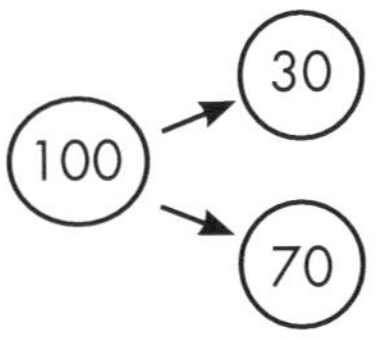

2. We can subtract tens from a 3-digit number mentally by subtracting 100 and then adding the extra tens back.

Subtract 60 from 849.
849 - 100 = 749
749 + 40 = 789

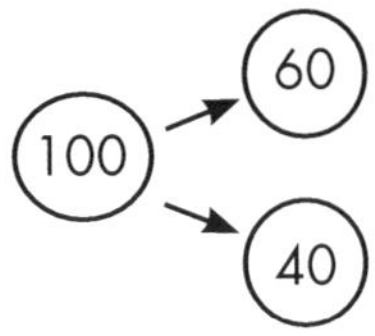

Concept:
Addition and subtraction involving 3-digit numbers and hundreds.

1. We can add hundreds to a 3-digit number mentally by using the "add hundreds" strategy.

Add 358 and 200.
300 + 200 = 500
500 + 58 = 558

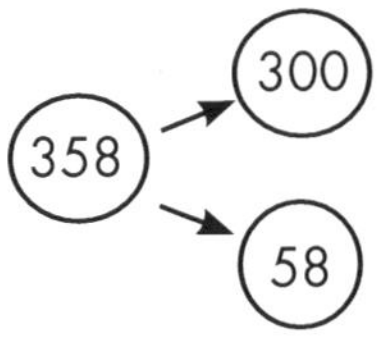

2. We can subtract hundreds from a 3-digit number mentally by using the "subtract hundreds" strategy.

Subtract 400 from 698.
600 - 400 = 200
200 + 98 = 298

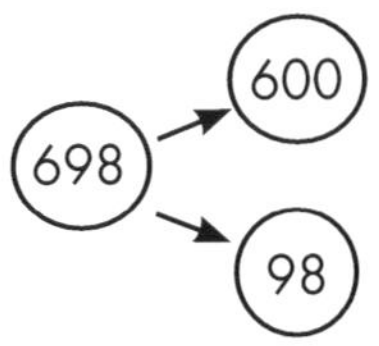

UNIT 9

Mental Calculations

POINTS
/50

Do the following sums mentally. (5 points each)

1. What is 38 + 7?

   10 → 7, 3

   38 + 10 = ☐

   ☐ – 3 = ☐

   So, 38 + 7 = ☐

2. What is 74 + 8?

   10 → 8, 2

   74 + 10 = ☐

   ☐ – 2 = ☐

   So, 74 + 8 = ☐

3. What is 66 + 5?

10 → 5, 5

66 + 10 = ____

____ – 5 = ____

So, 66 + 5 = ____

4. What is 77 + 15?

15 → 10, 5

77 + 10 = ____

____ + 5 = ____

So, 77 + 15 = ____

5. What is 48 + 17?

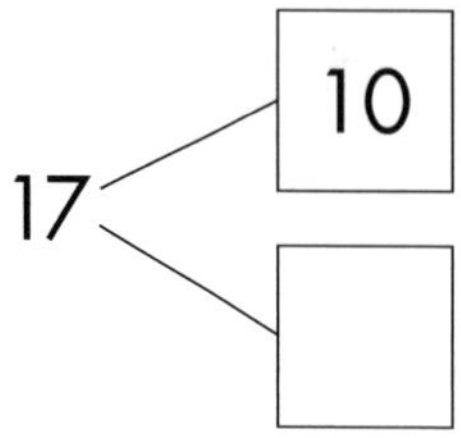

48 + 10 = ______

______ + ______ = ______

So, 48 + 17 = ______

6. What is 53 – 8?

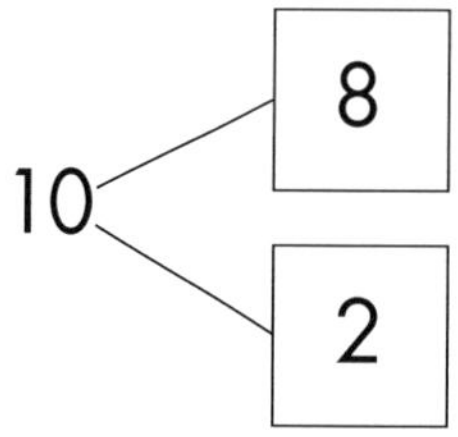

53 – 10 = ______

______ + 2 = ______

So, 53 – 8 = ______

7. What is 66 – 9?

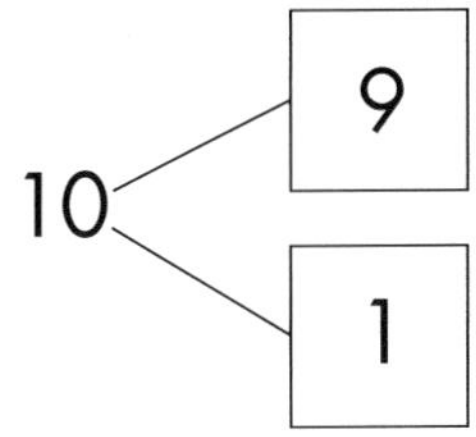

66 – 10 = ______

______ + 1 = ______

So, 66 – 9 = ______

8. What is 50 – 12?

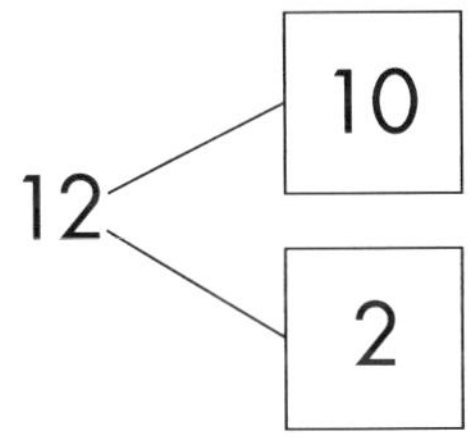

50 – 10 = ______

______ – 2 = ______

So, 50 – 12 = ______

9. What is 73 – 14?

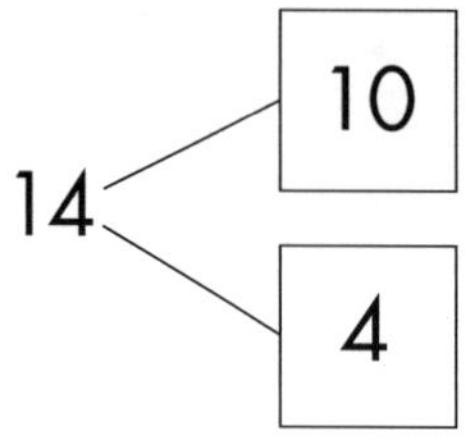

73 – 10 = ☐

☐ – 4 = ☐

So, 73 – 14 = ☐

10. What is 82 – 18?

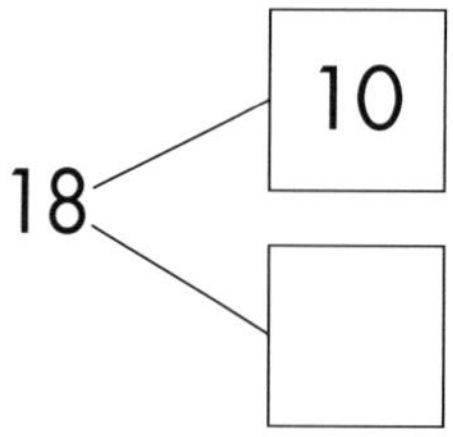

82 – 10 = ☐

☐ – ☐ = ☐

So, 82 – 18 = ☐

UNIT 10

Notes

Concept: Counting the amount of money in a given set of bills and coins.
We can count on to find out the amount of money we have.

We can count the dollars first and then the cents.

\$1 \$1 \$1 \$1 \$1 \$1

10¢ 10¢ 10¢ 10¢ 10¢ 10¢

+\$1 +\$1 +\$1 +\$1 +\$1

\$1, \$2, \$3, \$4, \$5, \$6

+20¢ +20¢

20¢, 40¢, 60¢

The total amount of money is \$6.60

Concept: Reading and writing money in decimal notation.
We can write money in dollars (\$) and cents (¢). We use "." to separate the dollars from the cents.

\$1 \$1 50¢

\$2.50 → cents (¢)
↓ dollar (\$)
→ dot

Two dollars and fifty cents.

Concept: Converting an amount of money in decimal notation to cents only and back to dollars.
We can change dollars to cents and cents to dollars.
\$1 = 100 ¢

\$1 \$1 \$1 \$1 \$1 \$1
\$1 25¢

\$7.25 = 725 ¢

25¢ 25¢ 25¢ 25¢
25¢ 25¢ 25¢ 25¢
25¢ 25¢

250 ¢ = \$2.50

We can add and subtract measures of money in the same way we do for whole numbers.
25¢ + 35¢ = 60¢
\$67 – \$56 = \$11

UNIT 10

Money

POINTS
/50

1. Color 3 coins to make 40 cents. (1 point)

   25¢ 25¢ 10¢ 10¢ 5¢ 5¢

2. Color 5 coins to make $0.75. (1 point)

   25¢ 25¢ 10¢ 10¢ 5¢ 5¢

3. Color 6 coins and three bills to make $3.66. (1 point)

   $1 $1 $1

   25¢ 25¢ 10¢ 10¢

   5¢ 5¢ 5¢ 1¢ 1¢

4. Match. (4 points)

(a)	25¢ 25¢ 10¢ 5¢	60¢
(b)	$1 10¢ 10¢ 1¢	$1.21
(c)	25¢ 25¢ 10¢	$6.35
(d)	$5 $1 25¢ 5¢ 5¢	65¢

5. Fill in the boxes. (4 points)

(a) 6 dollars 30 cents = $ [ ] $3.05 = [ ] ¢

(b) 2 dollars 35 cents = $ [ ] $0.45 = [ ] ¢

6. Fill in the blanks. (4 points)

(a) 125¢ = $ ________ $1.30 = ________ ten cents

(b) $2.00 = ________ dollars ________ cents

(c) Forty dollars and sixty cents = $ ________

7. Match the sums of money to the correct items. (4 points)

(a)	$10 $10 $5 $1
(b)	$10 $10 $1 $1
(c)	$10 $5 $1
(d)	$5 $5 25¢ 25¢ 25¢ 25¢

8. Write the correct amount in the box. (4 points)

25¢ 25¢ 10¢ 5¢ 5¢	
\$10 \$10 \$5 \$1 \$1	
\$10 \$1 25¢ 10¢ 10¢	
\$5 \$1 25¢ 10¢ 5¢	

Complete the following. (6 points)

9. \$9 + \$6 = \$ ________

10. 95¢ + 65¢ = ________ ¢

11. \$8 – \$5 = \$ ________

12. 70¢ – 15¢ = ________ ¢

13. \$6 + \$3 = \$ ________

14. 495¢ – 100¢ = ________ ¢

15.

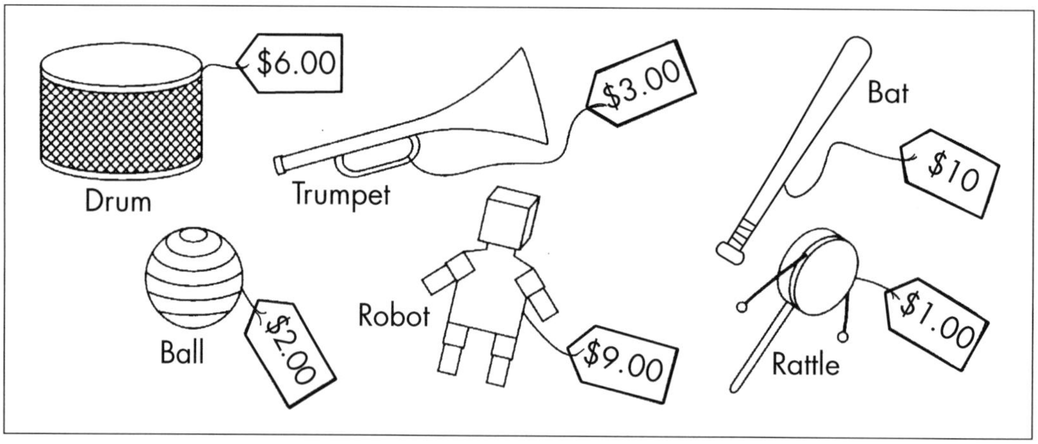

(a) David bought a ball and a rattle. (4 points)

They cost ________ altogether.

The bat costs ________ more than the trumpet.

(b) Ted had $10.

He bought a drum.

He had ________ left.

(c) Roger bought a rattle.

He had $1 left.

He had ________ at first.

Solve these problems.

16. Miriam had $8.

She bought a doll for $7.

How much money did she have left? (3 points)

She had ________ left.

17. Miguel bought a belt for $5 and a tie for $9.

How much more did the tie cost than the belt? (3 points)

The tie cost ________ more than the belt.

18. Tony bought a book for \$4 and 5 pencils for \$1.
How much did he spend altogether? (3 points)

He spent ________ altogether.

19. A teapot, a teacup, and a saucer cost \$12.
The teacup costs \$3.
The saucer costs \$2.
How much does the teapot cost? (4 points)

The teapot costs ________.

20. Carlos had $50.

He bought a handbag for his mother.

He used $12 to buy a dictionary and had $6 left.

How much did the handbag cost? (4 points)

UNIT 11 Notes

Concept: Telling and writing time to 5 minutes.

We can use a clock to help us tell time.

We read the time as 5 o'clock.
We write it as 5:00.

We read the time as five thirty.
We write it as 5:30.

The hour hand moves one complete round in 12 hours.

The minute hand moves one complete round in 60 minutes. There are 60 minutes in 1 hour.

Concept:
Use of hr. and min.
Duration of one hour/half hour from an o'clock time
We can use "hr." to stand for hour and "min." to stand for minutes.

1. What time is 1 hr. before 10 o'clock?

9 o'clock is 1 hr. before 10 o'clock.

2. What time is 30 min. after 8:00 o'clock?

8:30 is 30 min. after 8 o'clock.

Concept:
We use "a.m." to refer to time from midnight to just before noon and "p.m." from noon to just before midnight.

9 a.m.

7 p.m.

Concept:
We can draw hands on a clock face to show time.

1:30 p.m.

3:30 a.m.

8:00 p.m.

UNIT 11 Time

POINTS
50

1. Match the clocks to the correct time. (5 points)

3:15

11:55

12:05

7:10

6:50

Write the correct time in the blanks. (4 points)

2.  ____________

3.  ____________

4.  ____________

5.  ____________

Draw the minute hands on the clocks. (4 points)

6. 

8:25

7. 

4:40

Write the correct time in the blanks. (6 points)

8. 

Mother put a cake in the oven at ________.

9. 

A meeting began at ________.

10. 

The time is ________.

11.  (2 points)

The time is ________.

Complete the following. (12 points)

12. 

Mary went jogging at ________ p.m. in the evening.

It is ________ minutes to ________ o'clock.

It is ________ minutes past ________ o'clock.

13. 

Linda has a gym lesson at ________ a.m.

It is ________ minutes to ________ o'clock.

It is ________ minutes past ________ o'clock.

14. 

Danesh walks home from school at ________ p.m.

It is ________ minutes to ________ o'clock.

It is ________ minutes past ________ o'clock.

Draw the hour hand and the minute hand to show the time given in these exercises correctly. (8 points)

15. 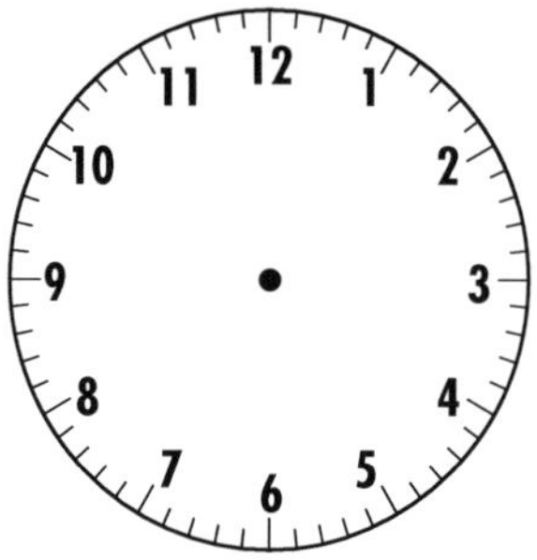

Paul helps his teacher in the library at 10:37 a.m.

16. 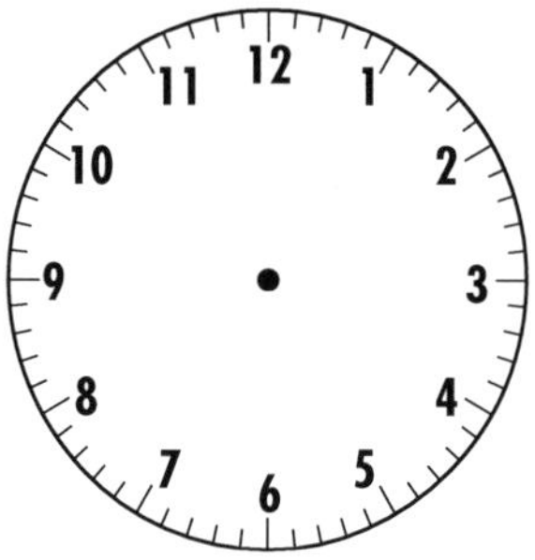

Jim eats his macaroni at 1:46 p.m.

17. 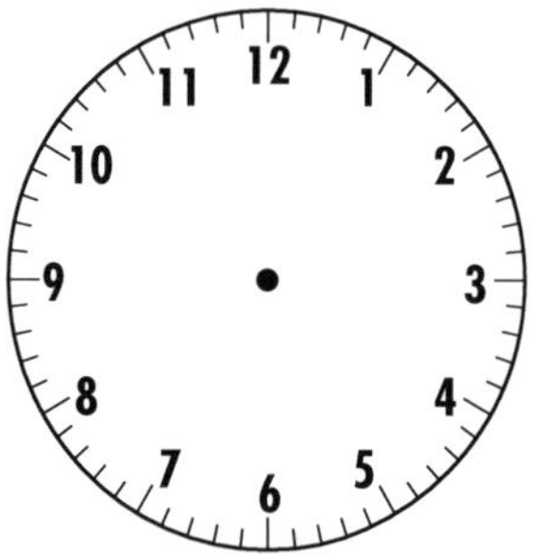

Alyssa reads her storybook at 8:08 p.m., after doing her homework.

18. 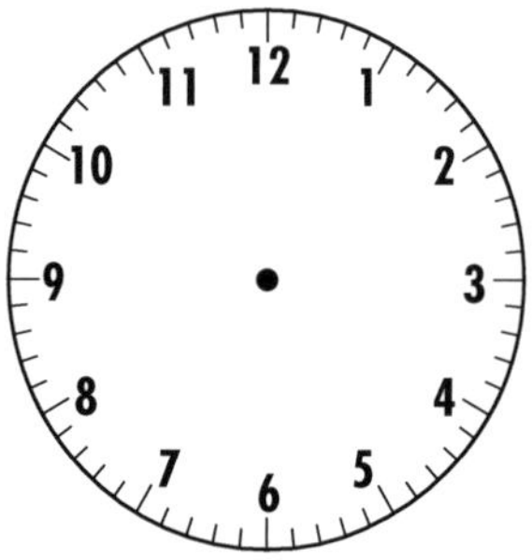

Mr. Baez tunes in a little late to his favorite TV series, at 9:12 p.m.

Draw the hands on the clocks and fill in the boxes with the correct answers.

19. Lily has to attend her ballet class at 10 a.m.
She needs to leave her home half an hour before 10 a.m.
At what time does she leave her house? (3 points)

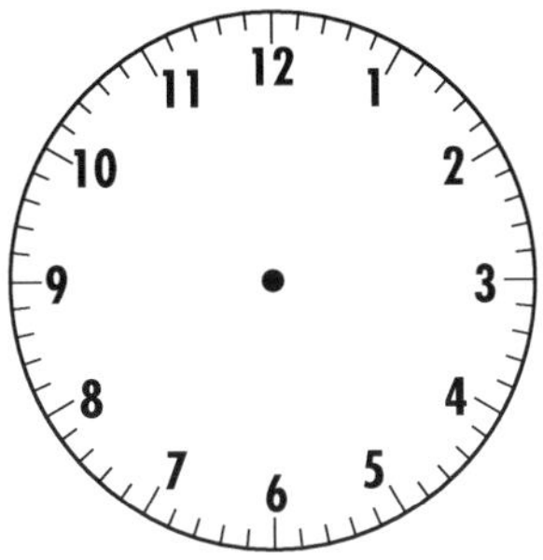

20. James read a novel for 40 minutes. He started reading at 10 p.m. At what time did he stop reading? (3 points)

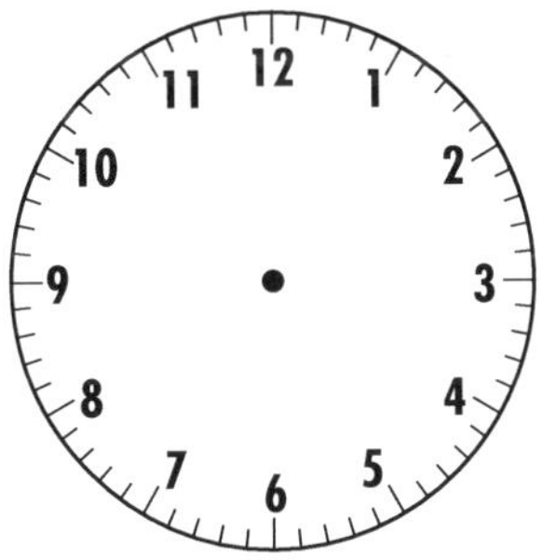

21. Raul's watch is half an hour slower than the school clock.
The school clock shows 2 p.m.
What is the time on Raul's watch? (3 points)

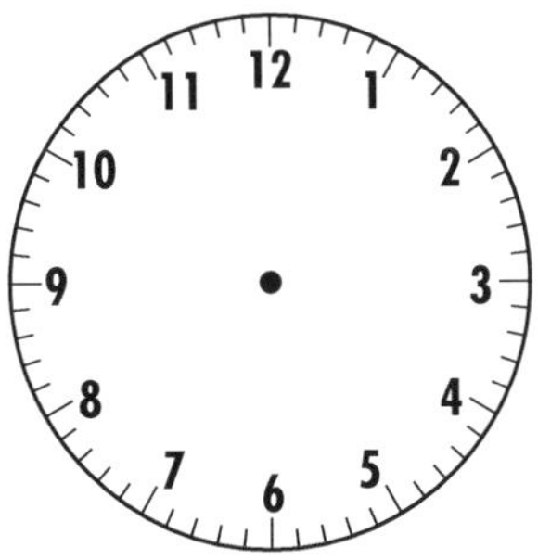

UNIT 12 Notes

Concept: Interpretation of fractions as part of a whole.
Reading and writing fractions.

When a whole is divided into a number of equal parts, each part is called a fraction.

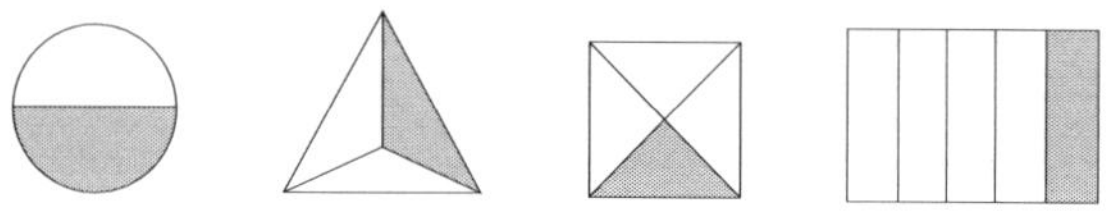

$\frac{1}{2}$	$\frac{1}{3}$	$\frac{1}{4}$	$\frac{1}{5}$
one-half	one-third	one-fourth	one-fifth

Concept:
Comparing and ordering – unit fractions, like fractions
Unit fractions are fractions where the top number of the fraction is 1.
For example, $\frac{1}{2}$, $\frac{1}{3}$, $\frac{1}{4}$, $\frac{1}{5}$, $\frac{1}{6}$

The larger the number of parts, the smaller the fraction.

$\frac{1}{2}$ x 2

$\frac{1}{3}$ x 3

$\frac{1}{4}$ x 4

$\frac{1}{5}$ x 5

$\frac{1}{6}$ x 6

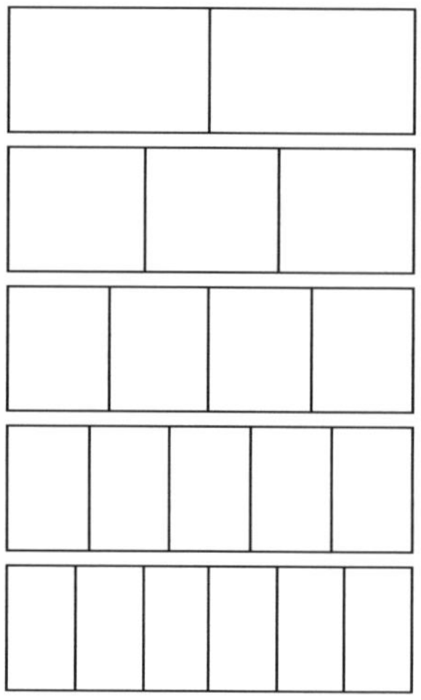

1. Which is larger, $\frac{1}{2}$ or $\frac{1}{3}$?

   $\frac{1}{2}$ is larger than $\frac{1}{3}$.

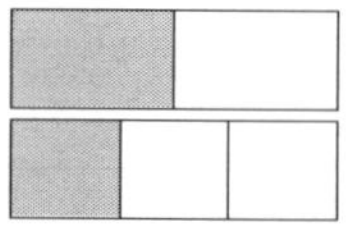

2. Which is smaller, $\frac{1}{5}$ or $\frac{1}{6}$?

   $\frac{1}{6}$ is smaller than $\frac{1}{5}$.

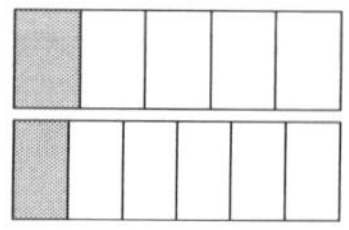

Like fractions are fractions where the bottom number of all the fractions are the same.
For example, $\frac{2}{4}$, $\frac{3}{4}$, $\frac{1}{4}$

We can add or subtract top numbers of like fractions in the same way we add or subtract whole numbers.
For example, $\frac{1}{7} + \frac{2}{7} = \frac{3}{7}$

For example, $\frac{8}{9} - \frac{4}{9} = \frac{4}{9}$

When we compare like fractions, the larger the top number, the larger the fraction. For example, Arrange the fractions $\frac{5}{9}$, $\frac{2}{9}$, and $\frac{7}{9}$ in order. Begin with the smallest.
$\frac{7}{9}$, $\frac{5}{9}$, $\frac{2}{9}$ → $\frac{2}{9}$, $\frac{5}{9}$, $\frac{7}{9}$

UNIT 12 Fractions

POINTS
/50

1. Circle the figures that are correctly shaded. (2 points)

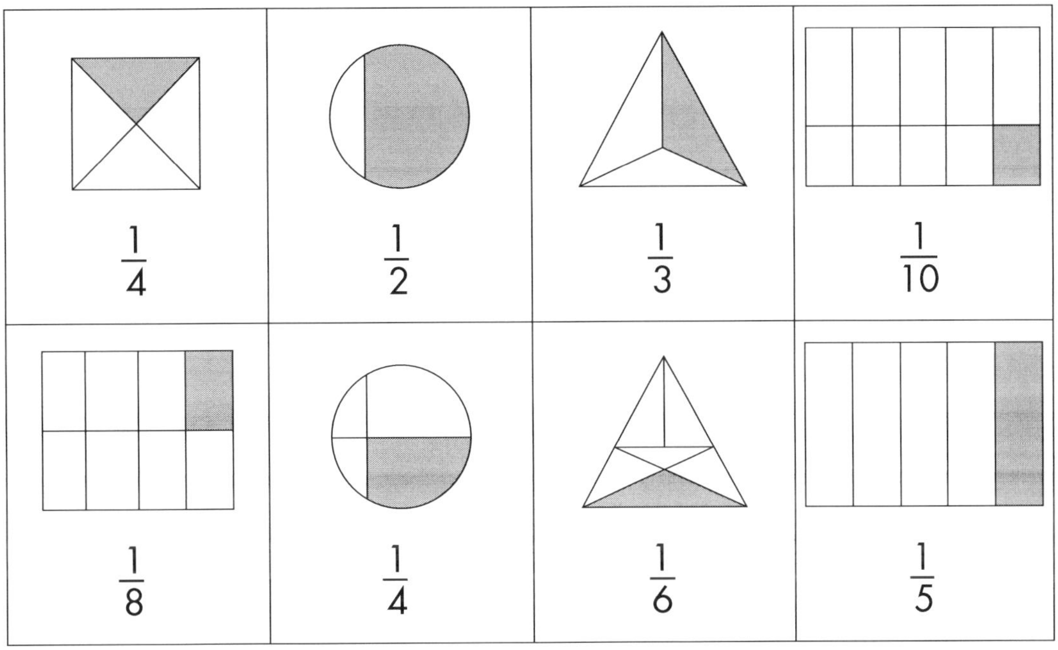

Color the bars to show each fraction.

2. $\frac{1}{3}$ $\frac{1}{6}$

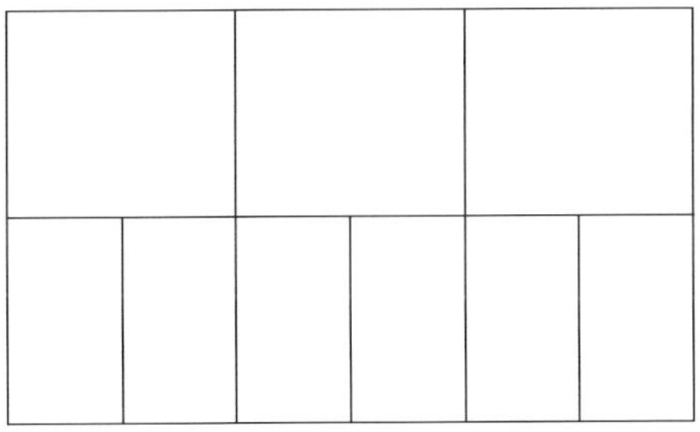

Which is the larger fraction?

________ (3 points)

3. $\frac{3}{4}$ $\frac{5}{8}$

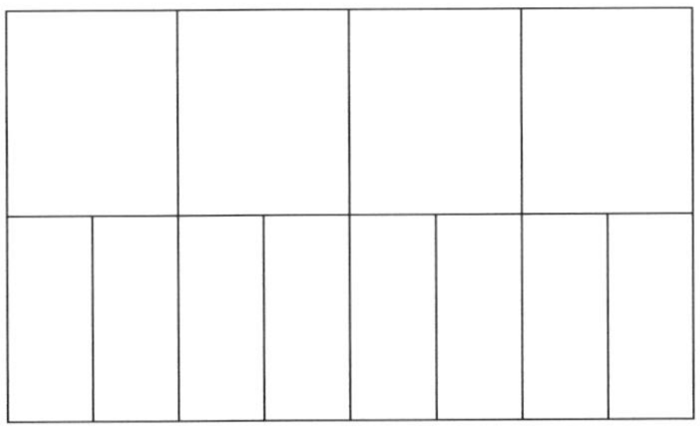

Which is the smaller fraction?

________ (3 points)

4. Color the boxes to show each of the given fractions.
Then, write the fractions in order.
Begin with the smallest fraction. (2 points)

$\frac{1}{2}$

$\frac{1}{6}$

$\frac{1}{4}$

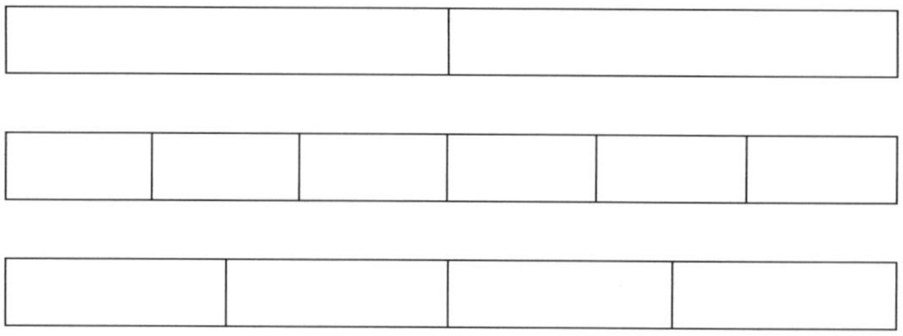

____

smallest

5. Color the boxes to show each of the given fractions.
Then write the fractions in order.
Begin with the largest fraction. (2 points)

$\frac{3}{8}$

$\frac{5}{6}$

$\frac{7}{12}$

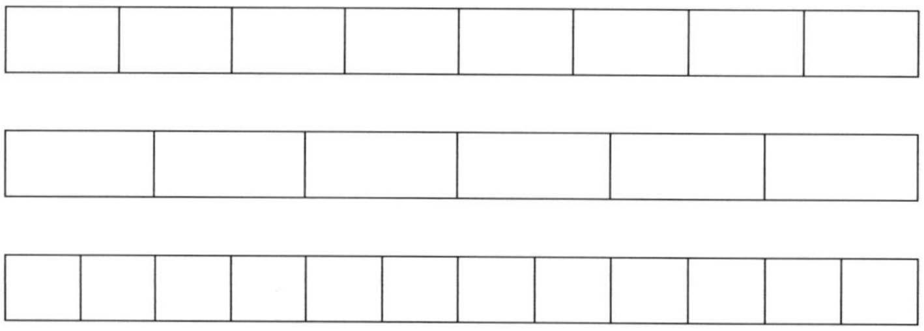

____

largest

Color the boxes to show each given fraction. (1 point each)

6. 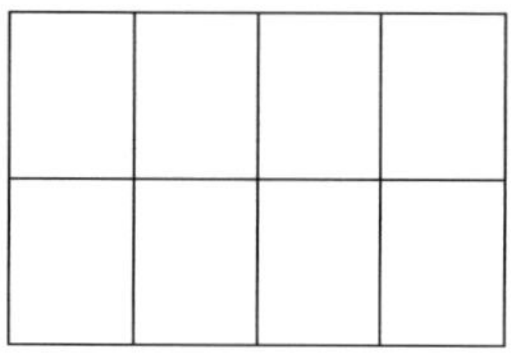

$\frac{5}{8}$

7. 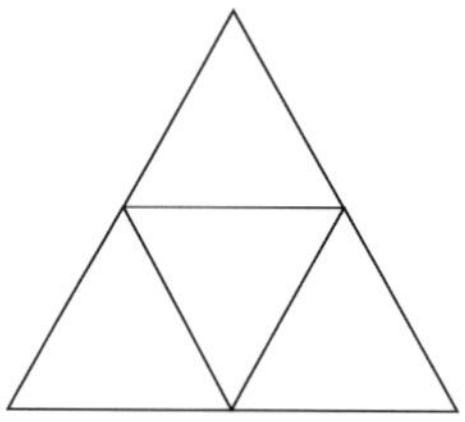

$\frac{3}{4}$

8. 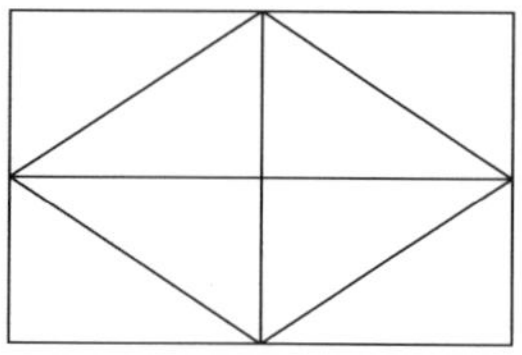

$\frac{3}{8}$

9. Each shaded triangle shows a given fraction.
Match each triangle to the correct fraction shown. (4 points)

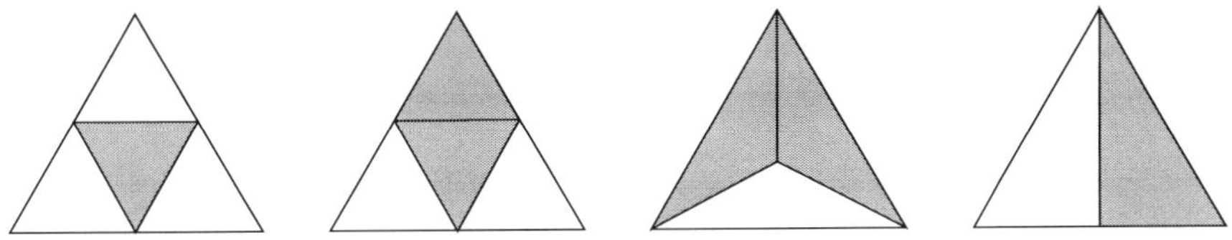

$\frac{2}{3}$ $\frac{1}{2}$ $\frac{1}{4}$ $\frac{2}{4}$

Write the correct answers in the boxes.

10. 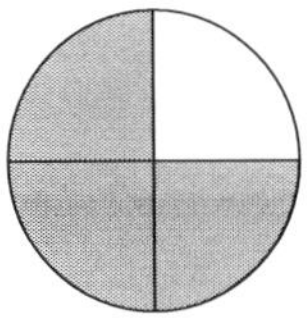

☐ out of 4 equal parts are shaded.

☐ of the circle is shaded. (2 points)

11. 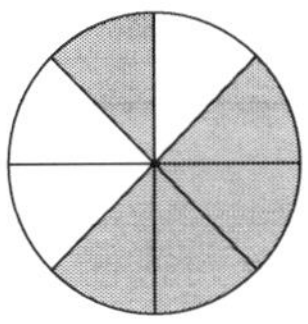

☐ out of ☐ equal parts are shaded.

☐ of the circle is shaded. (6 points)

Write the correct fraction for each shaded figure. (1 point each)

12. 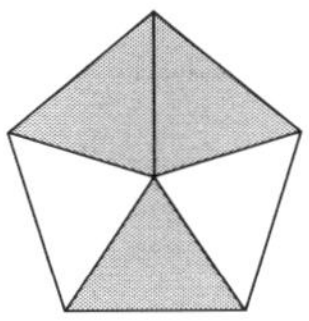

______

13. 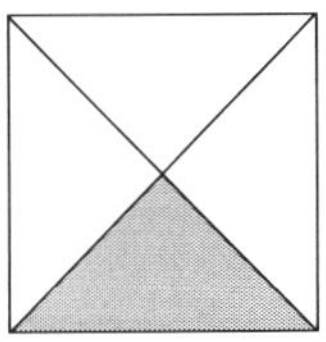

______

14. 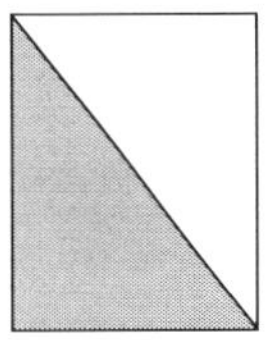

______

15. 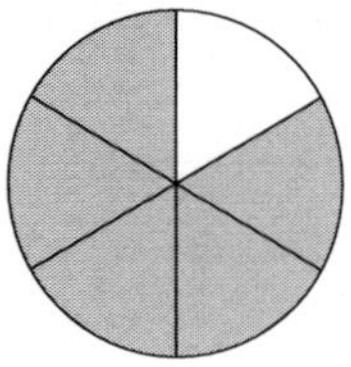

____

16. 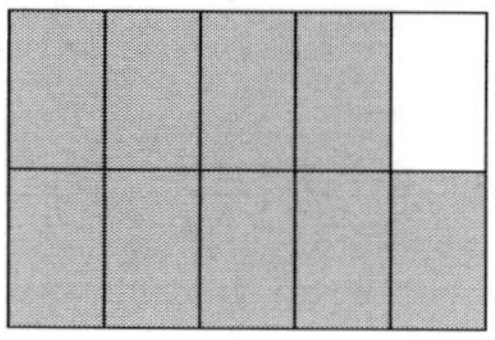

____

17. 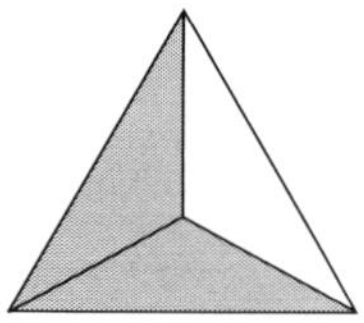

____

18. 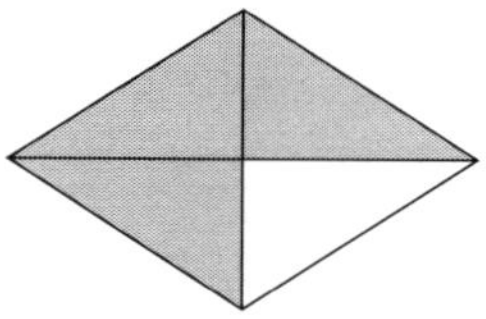

____

Write answers to the following questions. (3 points)

19. 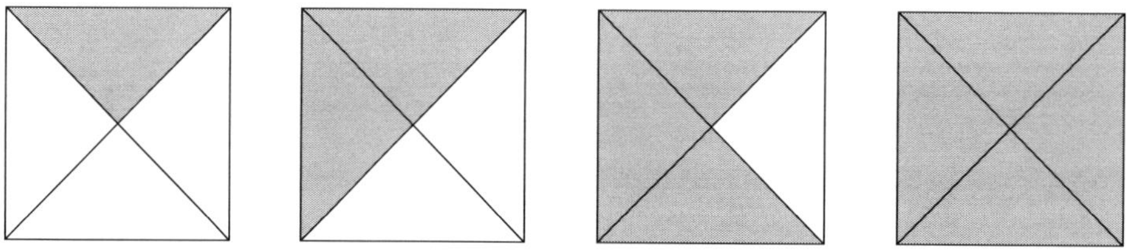

$\frac{1}{4}$ $\frac{2}{4}$ __________ __________

There are __________ quarters in 1 whole.

20. Color. Then fill in the blank. (3 points)

$\frac{12}{12}$ 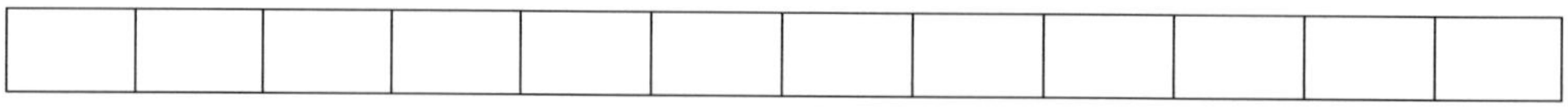

$\frac{5}{12}$ 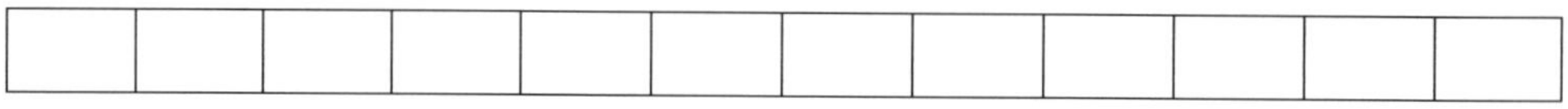

$\frac{5}{12}$ and __________ make 1 whole.

Circle the fraction that shows the shaded part of each figure.

(1 point each)

21.

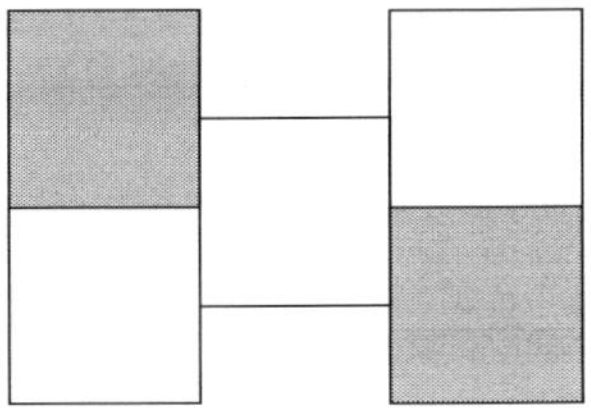

$\frac{2}{5}$ $\frac{2}{4}$ $\frac{3}{5}$

22.

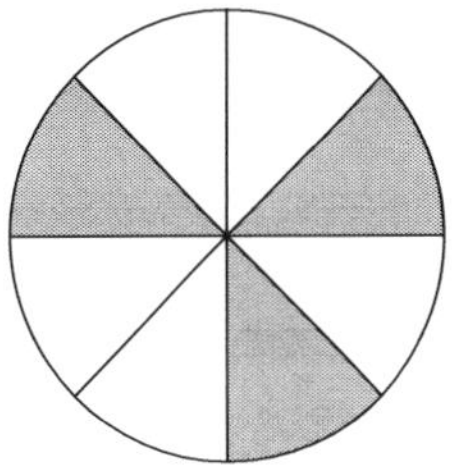

$\frac{1}{3}$ $\frac{3}{5}$ $\frac{3}{8}$

23.

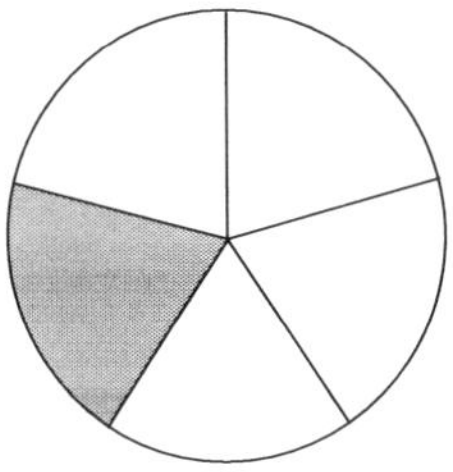

$\frac{1}{3}$ $\frac{1}{4}$ $\frac{1}{5}$

24.

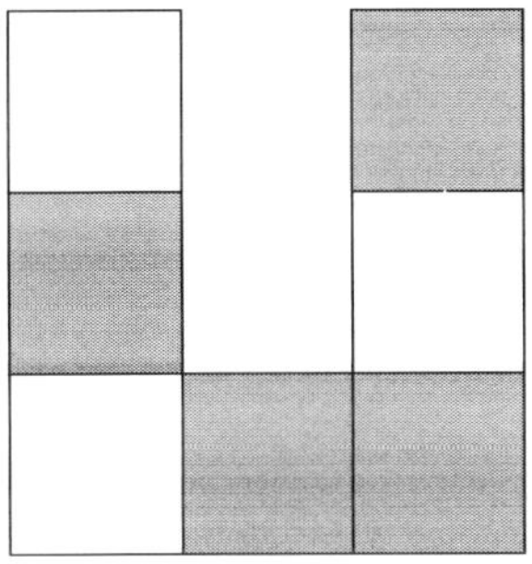

$\frac{3}{4}$ $\frac{4}{7}$ $\frac{4}{6}$

25. Write two fractions that are greater than $\frac{2}{3}$.

Use the following bars to help you. (4 points)

$\frac{2}{3}$

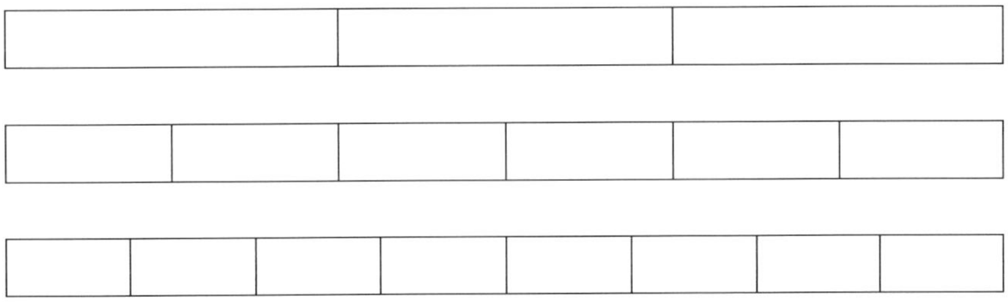

Two fractions that are greater than $\frac{2}{3}$ are

____

26. Write two fractions that are smaller than $\frac{3}{4}$. Use the following bars to help you. (4 points)

$\frac{3}{4}$

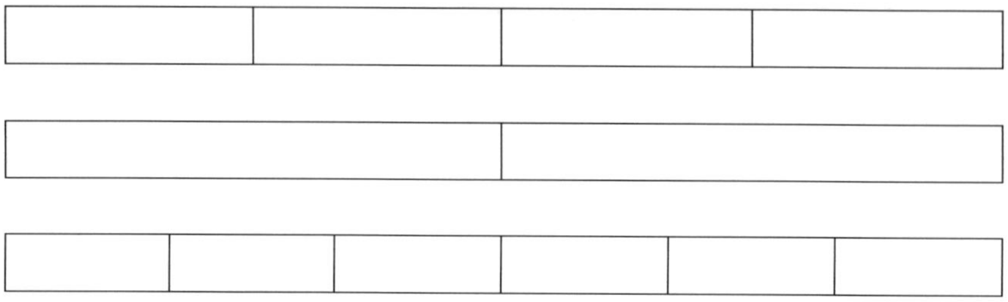

Two fractions that are smaller than $\frac{3}{4}$ are

____

27. What fraction of the figure is shaded? (4 points)

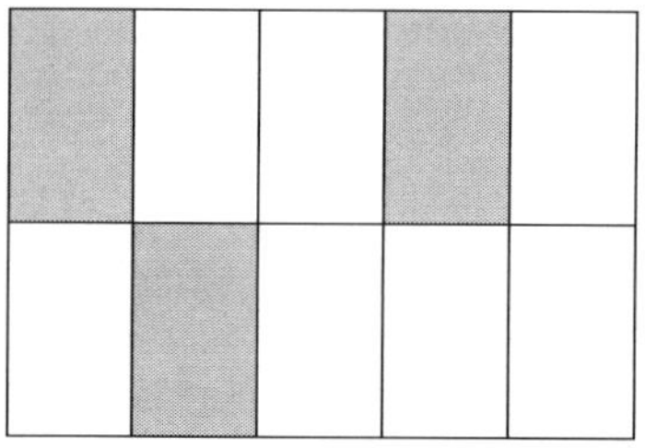

__________ of the figure is shaded.

Color more rectangles to show $\frac{1}{2}$ of the figure shaded.

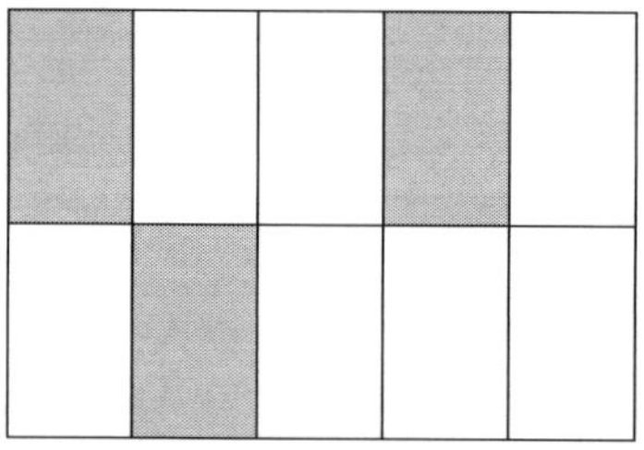

UNIT 13

Notes

Concept: Line, Curve, and Surface

This is a line. We can draw lines using rulers.

This is a curve. We do not need rulers to draw lines.

This is a surface.

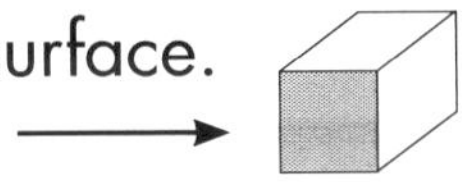

We can use lines and curves to draw figures and surfaces.

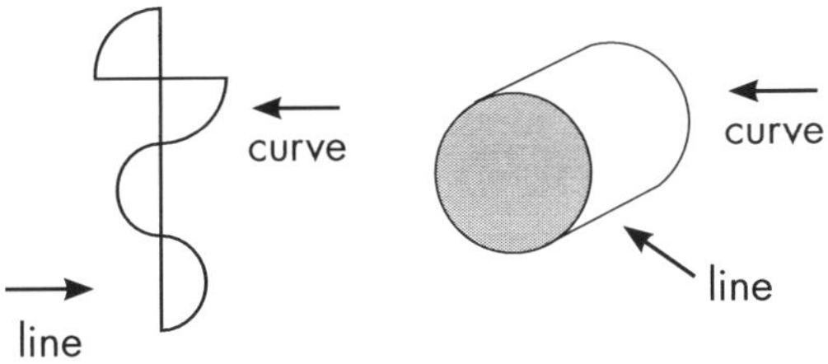

Concept:
Shapes and two-dimensional figures.

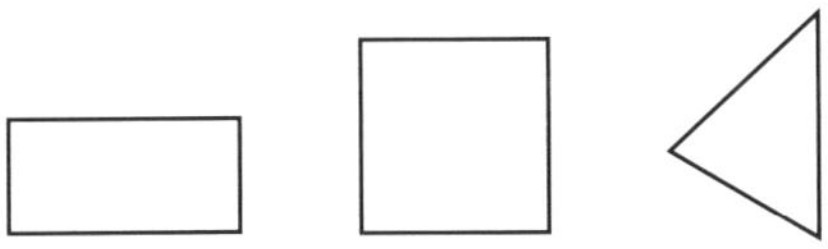

rectangle square triangle

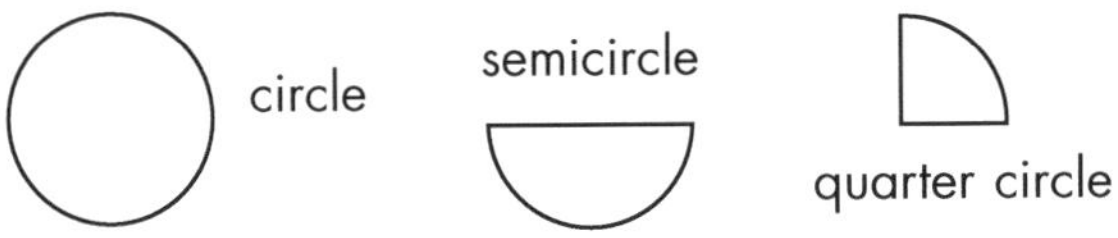

We can form different two-dimensional figures using different shapes.

Concept:
Shapes and three-dimensional figures.

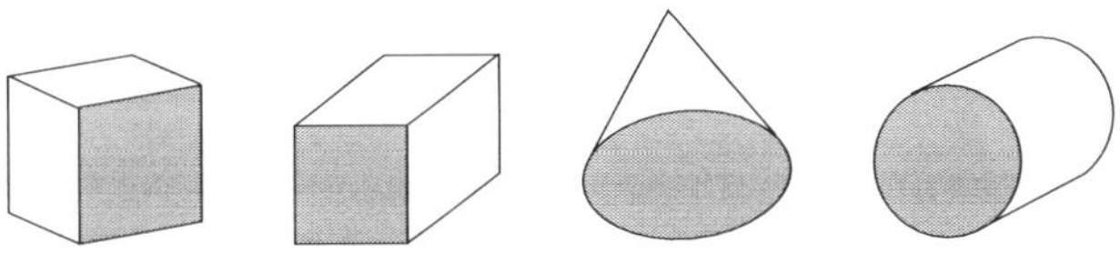

cube cuboid cone cylinder

We can form different three-dimensional figures using different shapes.

Concept:
We can make a pattern using 2-D figures or 3-D solids of different shape, size, color, and direction.

For example:

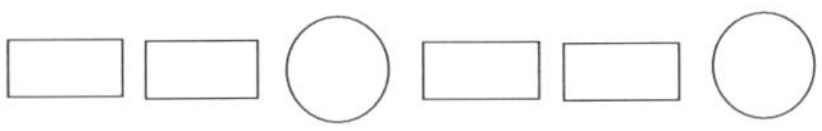

For example:

UNIT

13 Lines, Surfaces, Shapes, and Patterns

POINTS
/50

Look at the picture carefully and answer the following questions.

1. How many curves are there? ________ (1 point)
2. How many straight lines are there? ________ (1 point)
3. Draw a square on the flag. (1 point)
4. A square has ________ straight lines. (1 point)
5. How many straight lines are there now? ________ (1 point)
6. 

The picture has ________ straight lines and ________ curves. The hat is made up of ________ straight lines. There are ________ lines and curves altogether. (4 points)

Fill in the blanks with the correct answers.

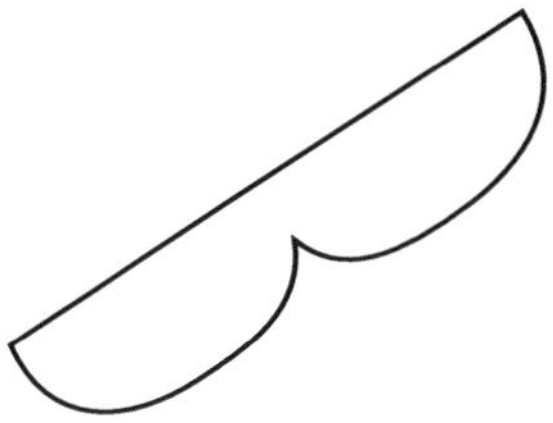

7. There are ________ straight lines and ________ curves.

(1 point)

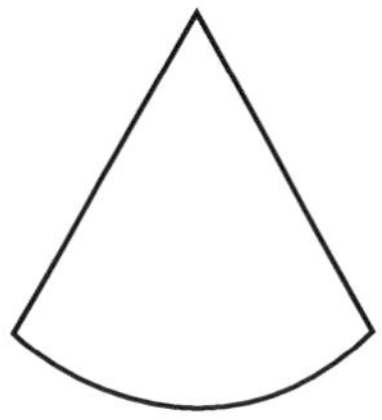

8. There are ________ straight lines and ________ curves.

(1 point)

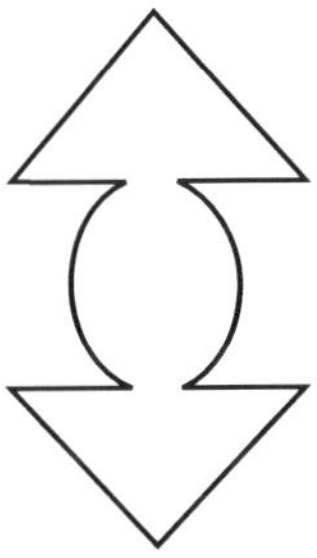

9. There are ________ straight lines and ________ curves.

(1 point)

10. There are ________ straight lines and ________ curves.

(1 point)

11. 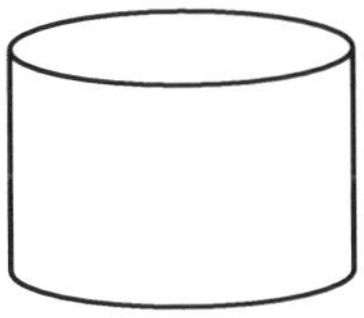

This solid has ________ curved face(s) and ________ flat face(s). (1 point)

12. 

In this figure there are ________ straight lines and ________ curves. (1 point)

Write **curved** or **flat** in the box.

13. 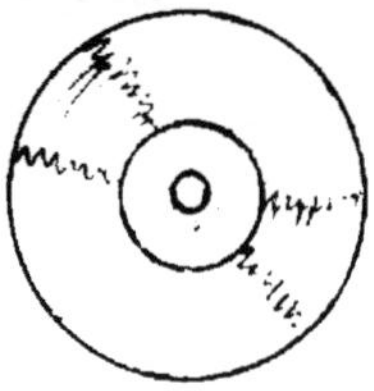

The disc has a ________ surface. (1 point)

14. 

The letter has a ________ surface. (1 point)

15. 

The ball has a ________ surface. (1 point)

Circle the two objects that have the same shape. (1 point each)

16. 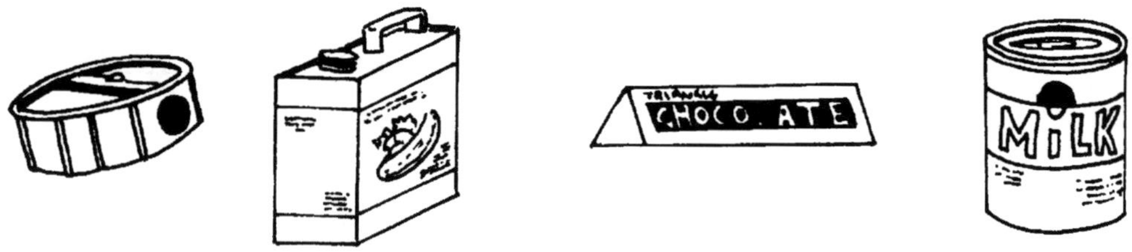

17. 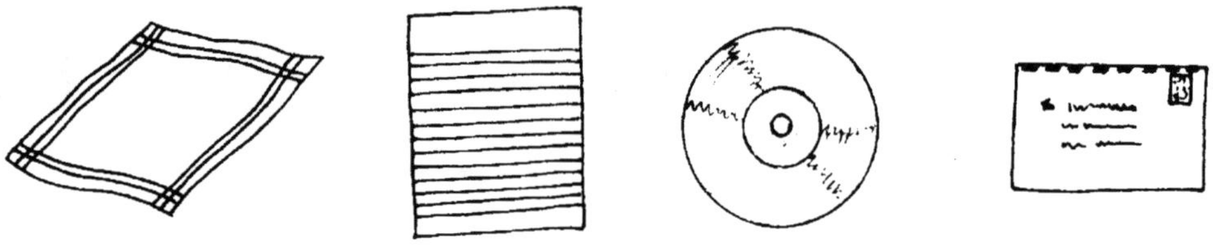

Write the name of each of the following shapes. (1 point each)

18. 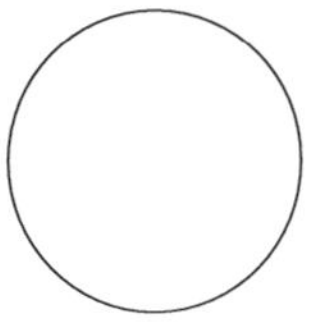

19. 

20. 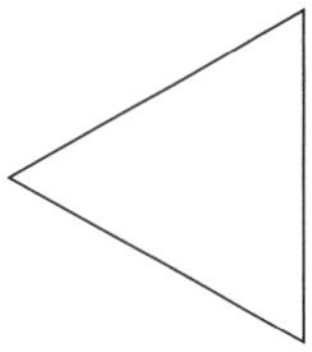

21. 

______ ______ ______ ______

22. Draw lines to join the shaded faces with the correct shapes. (4 points)

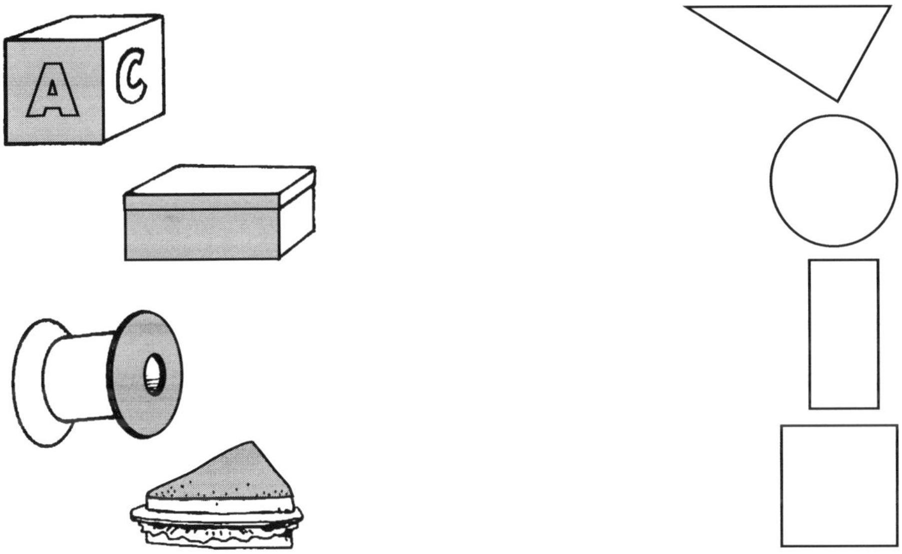

23. Name the shape of each of the following shaded faces. (4 points)

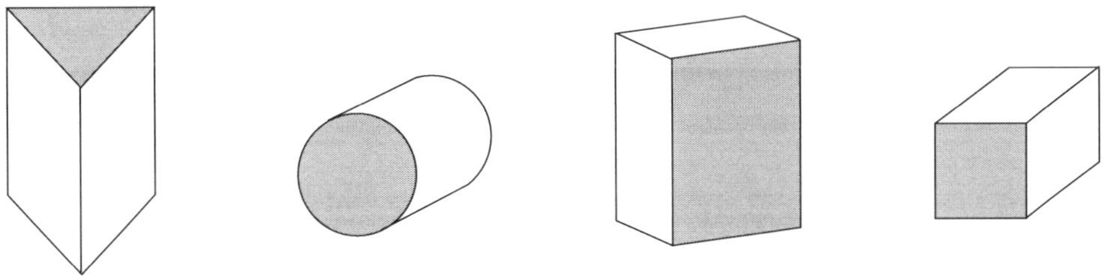

______ ______ ______ ______

24. Fill in the blanks with the correct answers. (1 point)

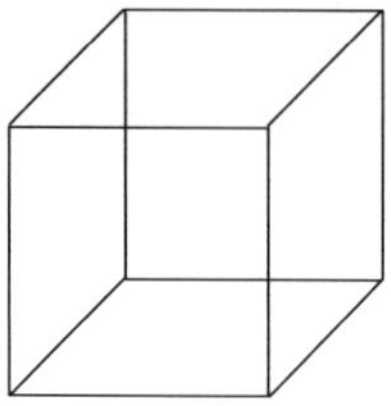	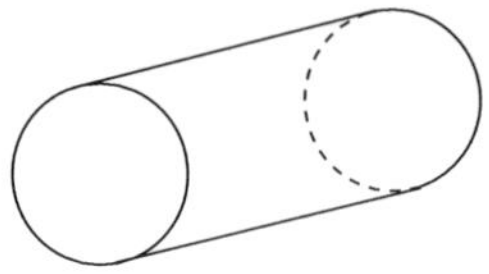
________ faces are squares.	________ faces are circles.

25. Study each pattern.

What comes next?

(a)

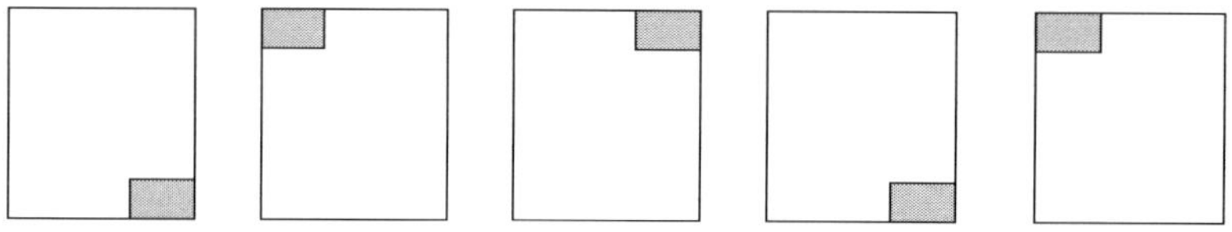

____

(1 point)

(b)

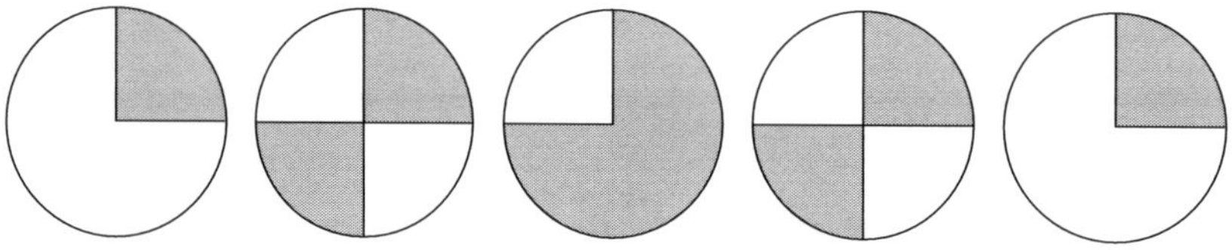

____

(1 point)

(c)

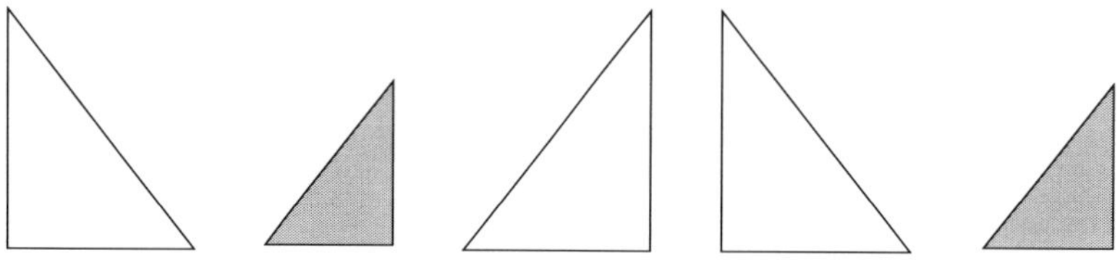

____

(1 point)

(d)

 ______

(1 point)

26. This figure is made up of three shapes.
Draw lines to show them.
Then name them. (3 points)

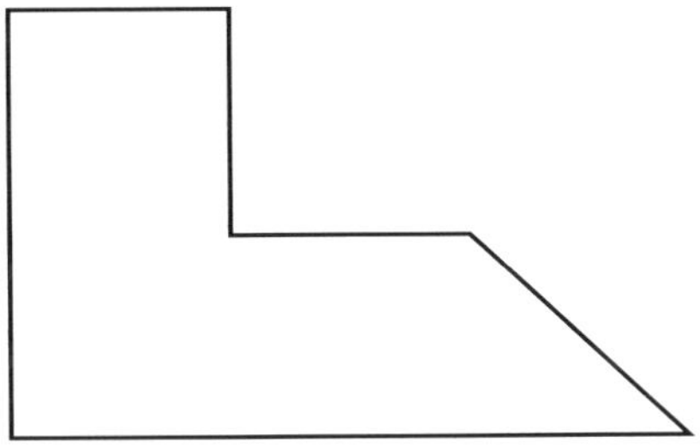

(a) ______
(b) ______
(c) ______

27. This figure is made up of 6 different shapes.
Draw lines to show them.
Then color the
(a) quarter-circles red,
(b) semicircle blue,
(c) triangle green,
(d) square yellow,
(e) rectangle brown.

(5 points)

28. Color the faces that make up the solid as shown on the left. (1 point)

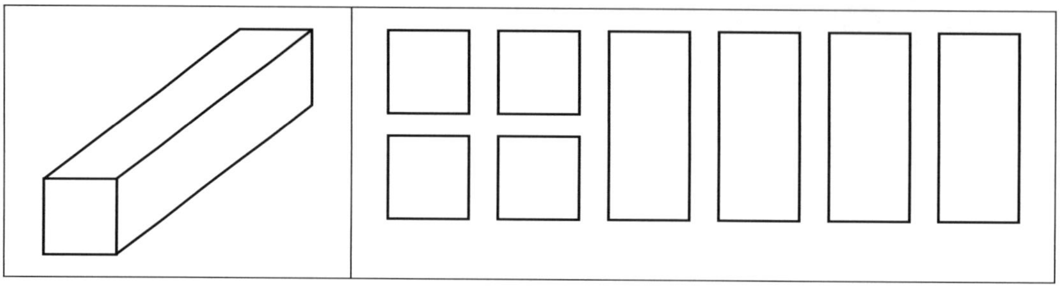

29. Color the next shape in the pattern. (2 points)

30. Color the missing part of the first shape in each row. (2 points)

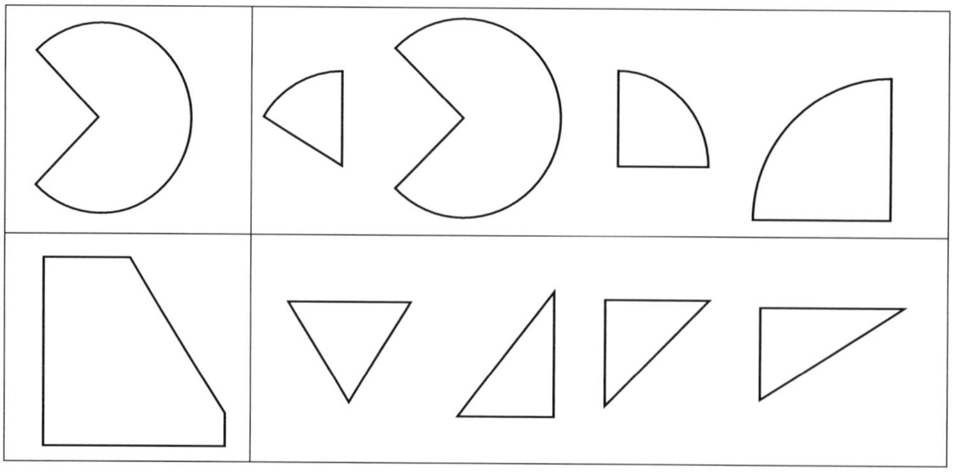

UNIT 14

Notes

Concept: We can present information using picture graphs.

1. Count each type of object shown.

The picture graph below shows the number of each type of toy.

Concept:
We can create picture graphs by using information recorded on a table or chart.

1. Susan collects postcards.
She counts the number of postcards she collected from January to May and records it in a table.

Month	Jan.	Feb.	Mar.	Apr.	May
Number	16	8	12	4	6

She then draws a picture graph to show the number of postcards she collected for each month.

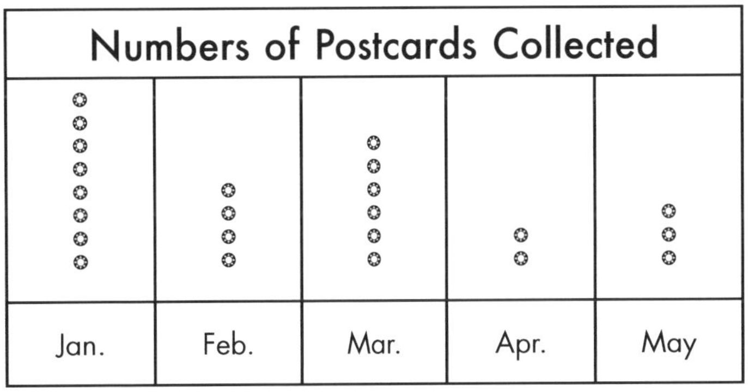

Each ○ stands for 2 postcards

UNIT 14

Picture Graphs

POINTS
/50

Complete the following sentences.

1. Each  stands for 5 cups. (4 points)

   (a)  stand for ________ cups.

   (b)  stand for ________ cups.

2. Each △ stands for 2 houses.
   Draw the correct number of △ to show: (4 points)

   (a) ____________________ stand for 4 houses.

   (b) ____________________ stand for 8 houses.

3. Study the graph and complete the table below. (8 points)

Pencils Sold in Our Bookshop	
Monday	
Wednesday	
Friday	
Each  stands for 4 pencils.	

Day	Monday	Wednesday	Friday	Total
No. of pencils sold				

4. Study the graph below.

Then, complete the following sentences.

Number of Fruit Trees in Mr. Martin's Garden			
△ △ △	△ △	△ △ △ △ △	△
Apple trees	Peach trees	Pear trees	Cherry trees

Each △ stands for 3 trees.

(a) Mr. Martin has ________ peach trees. (1 point)

(b) He has 9 ________ trees. (1 point)

(c) He has ________ more apple trees than cherry trees. (1 point)

(d) He has ________ fewer peach trees than apple trees. (1 point)

(e) The total number of ________ trees and ________ trees is the same as the number of pear trees. (2 points)

(f) He has ________ fruit trees altogether. (2 points)

5. Sally drew the animals she saw in a pet shop.

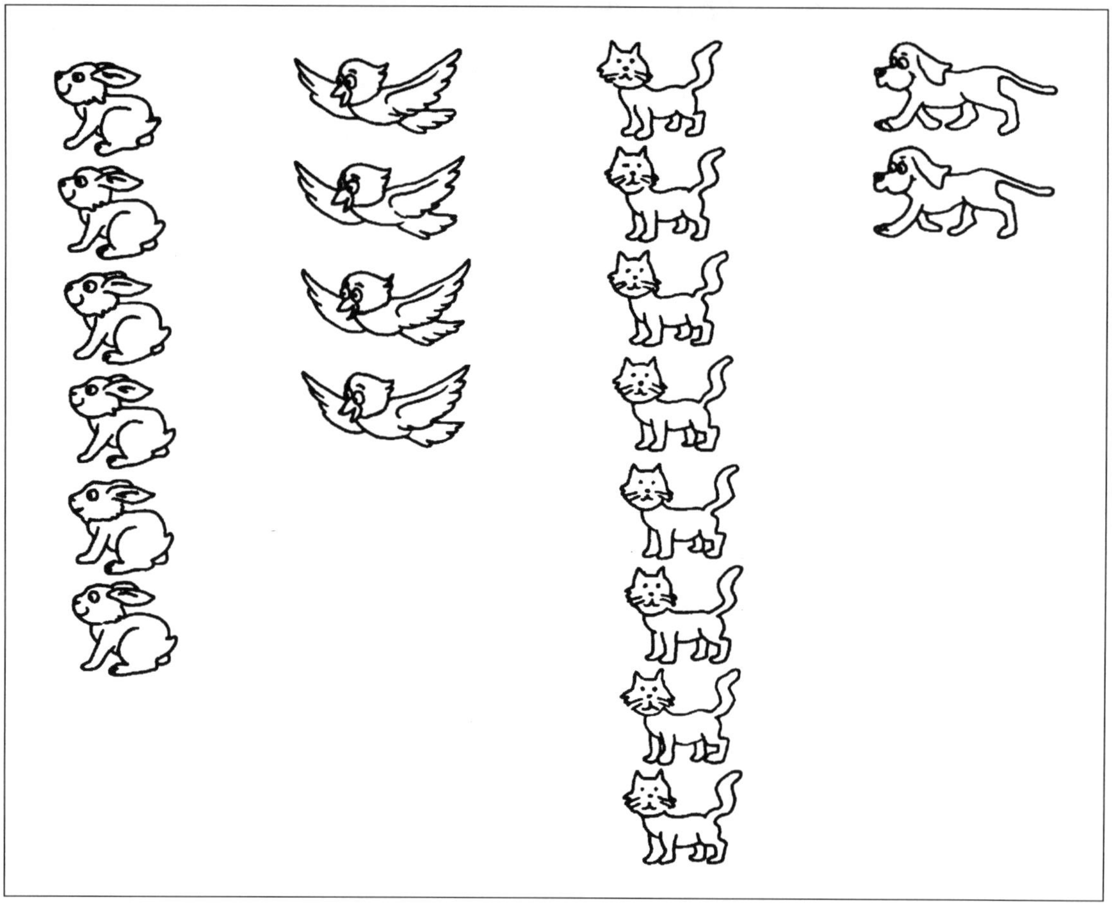

Help Sally complete the following graph. (6 points)

Animals in a Pet Shop			
▲			
▲			
▲			
Rabbits	Birds	Cats	Dogs
Each ▲ stands for 2 animals.			

The graph shows the weights of four animals.
Study it and fill in the blanks.

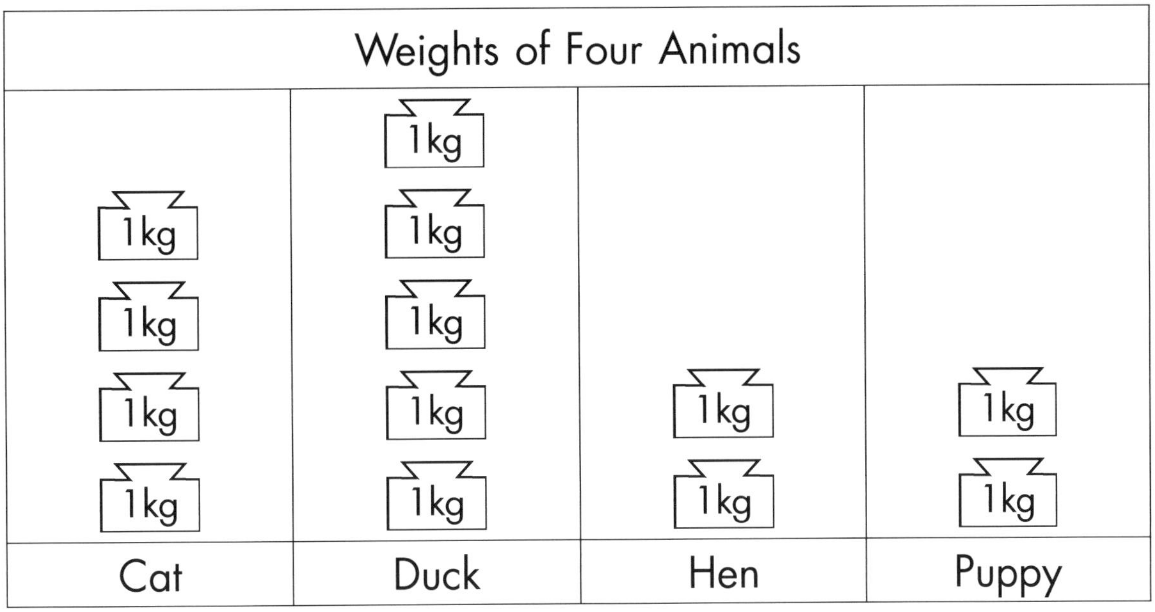

6. (a) The weight of the cat is about ______ kilograms. (1 point)

   (b) The weight of the duck is about ______ kilograms. (1 point)

   (c) The weight of the puppy is about ______ kilograms. (1 point)

   (d) The hen is as heavy as the ______. (1 point)

   (e) The ______ is heavier than the cat. (1 point)

   (f) The duck is ______ kg more than the hen. (1 point)

   (g) The puppy is ______ kg lighter than the cat. (2 points)

   (h) The weight of the cat is the same as the total of the weight of the ______ and the ______. (2 points)

   (i) The weight of ______ puppies is heavier than a duck. (2 points)

7. The picture graph below shows the number of students in each classroom who took part in a concert.

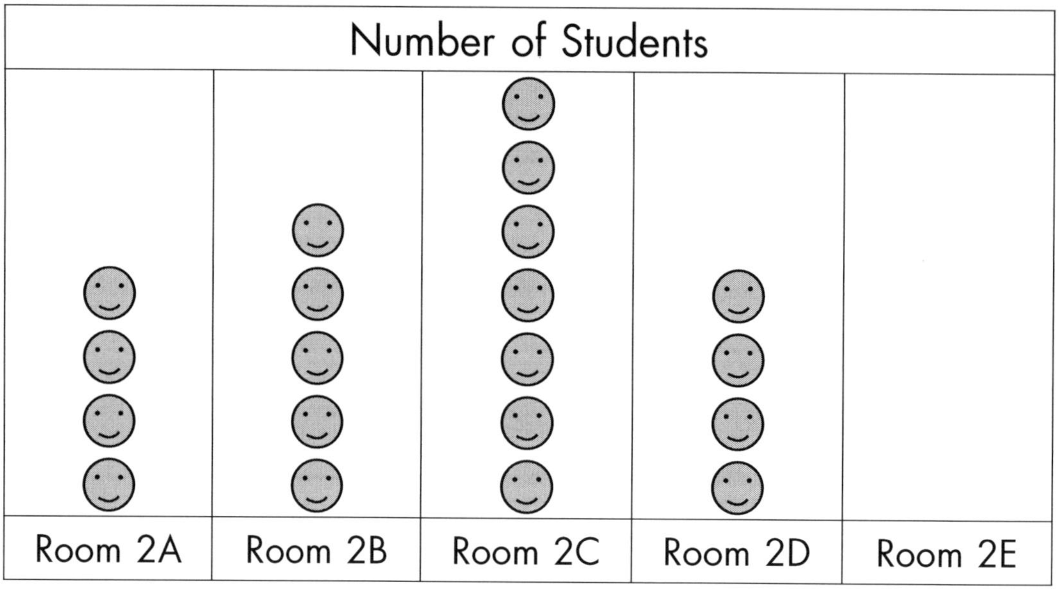

(a) The same number of students in Room ________ and Room ________ took part in the concert. (2 points)

(b) In Room 2B, 20 students took part in the concert. Each  stands for ________ students. (2 points)

(c) 12 fewer students in Room 2A took part in the concert than in Room ________. (2 points)

(d) To have the same number of students in both Room 2C and 2D take part in the concert, ________ more students must take part in 2D. (2 points)

(e) A total of 100 students took part in the concert. ________ students in Room 2E took part in the concert. Complete the graph. (3 points)

(f) In Room 2E, 10 boys and ________ girls took part in the concert. (2 points)

UNITS 1 to 14

Test Yourself 1

POINTS ___ / 100

Section A (80 points)

1. Count and write the correct numbers in the boxes. (2 points)

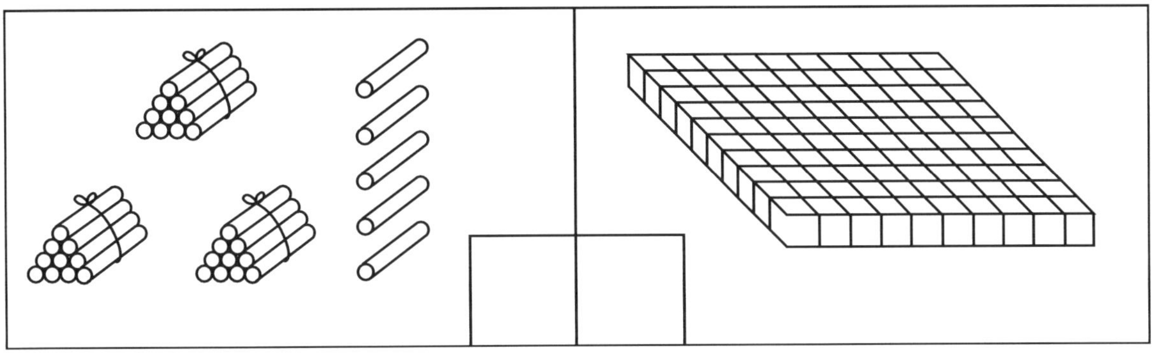

2. Write the following statements in numbers. (3 points)

   Five hundred and eight ________

   Six hundred and twelve ________

   Four hundred and two ________

3. Circle the smallest number in each set. (2 points)

35	53	51

129	200	140

4. The value of the digit 6 in 608 is ________. (1 point)

5. The digit 7 in the number of 273 stands for ☐ × 10 (1 point)

6. Given 3 hundreds + 4 tens + 2 ones, write the result in words. (1 point)

______________________________.

7. Complete: 10, 20, 30, ______, ______, ______, 70, 80, ______, ______. (1 point)

8. Complete the following number sentences. (4 points)

☐ − 4 = 7

27 + 4 = ☐

200 + ☐ = 1000

407 + 98 = ☐

9. Given 19, 40, and 21, write two subtraction sentences with them. (4 points)

☐ ○ ☐ = ☐

☐ ○ ☐ = ☐

10.

The nail is about ______ centimeters long. (1 point)

11. Jane walked to the bank and then to the coffee shop to have her lunch. How far did she walk? (2 points)

She walked ________ meters.

12. Study the picture. Then fill in the blank. (1 point)

The girl has a weight of about ________ kilograms.

13. Study the pictures. Then fill in the blanks. (2 points)

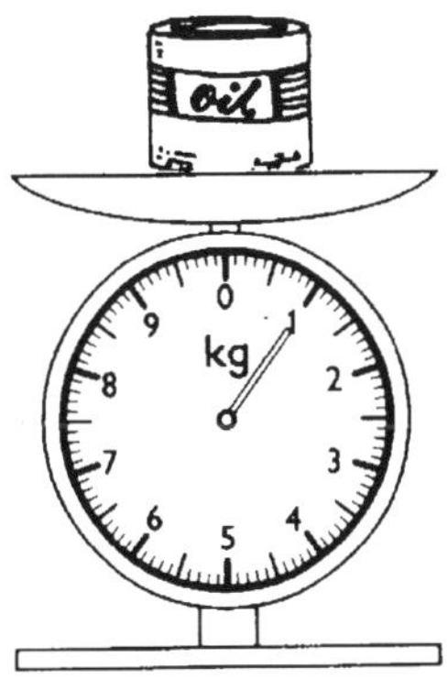

The can of oil has a weight of about ________ kilograms.

The can of oil has a weight of about ________ kilograms less than the bag of sugar.

14. Write two multiplication sentences for this picture. (4 points)

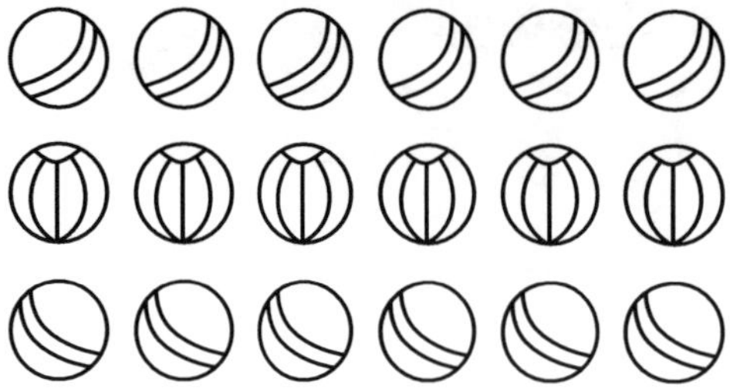

☐ × ☐ = ☐

☐ × ☐ = ☐

Fill in the blanks. (7 points)

15. 1 cat has 4 legs.

3 cats have ________ legs.

5 cats have ________ legs.

16. 3 × 5 = ________ 3 × 10 = ________

17.

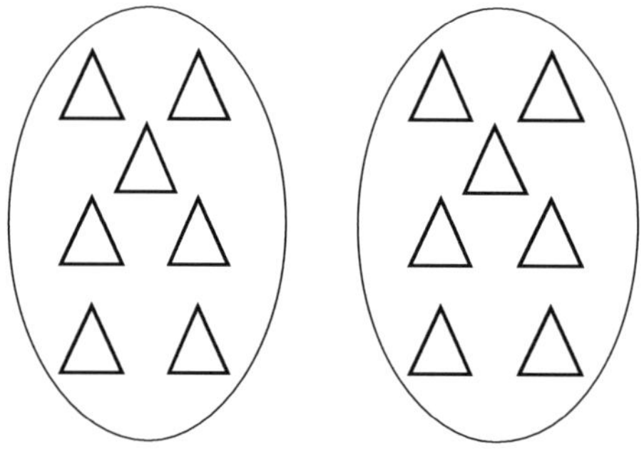

2 × ________ = 14

14 ÷ 2 = ________

18. How many fives are there in 40? ________

19. Each 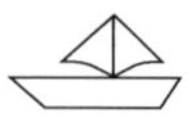 stands for 2 boats.

Draw the correct number of boats for the following statements.

________________________ stand for 8 boats. (1 point)

________________________ stand for 6 boats. (1 point)

20. Complete the following. (3 points)

______ × 4 = ______

There are ______ apples altogether.

21. Divide 28 tops into groups of 4. (2 points)

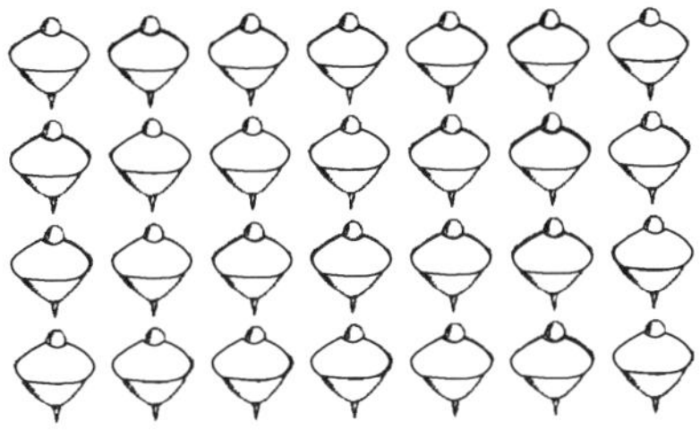

28 ÷ 4 = ______

Fill in the missing numbers. (6 points)

22. 8 × 5 = $\square$

8 × 5 is 5 more than ______ × 5.

23. 6 × 10 = $\square$

6 × 10 is 10 more than ______ × 10.

24. 9 × 4 = $\square$

9 × 4 is 4 less than ______ × 4.

25. $\frac{1}{12}$ and ______ make 1 whole. (2 points)

$\frac{12}{12}$

$\frac{1}{12}$

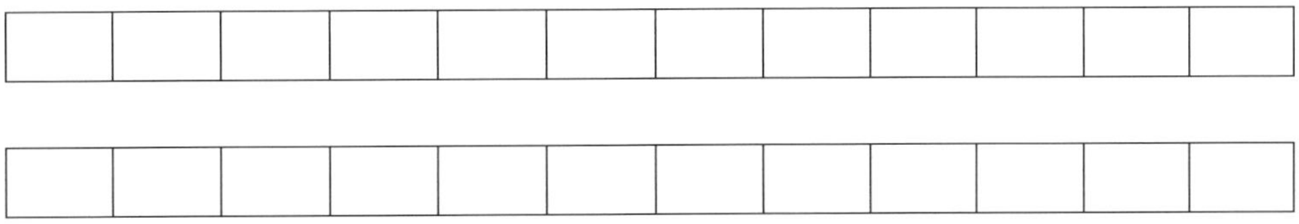

26. Circle the fractions that are the same as 1 whole. (2 points)

$\frac{6}{6}$ $\frac{5}{7}$ $\frac{9}{9}$ $\frac{3}{4}$

27. Circle the largest fraction. (2 points)

$\frac{1}{8}$ $\frac{1}{2}$ $\frac{1}{3}$ $\frac{1}{6}$

28. The fraction stands for the shaded part of each figure. Circle the figure that does **not** match the fraction. (2 points)

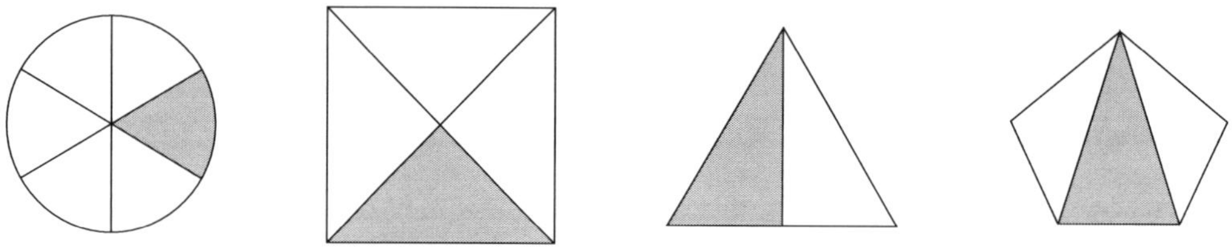

$\frac{1}{6}$ $\frac{1}{4}$ $\frac{1}{2}$ $\frac{1}{3}$

29. 

3 o'clock in the morning is ________. (2 points)

3 o'clock in the afternoon is ________. (2 points)

30. (a) Jason gets 4 quarts of water each time he draws from the barrel.

How many times must he draw to get 16 quarts of water? ________ times

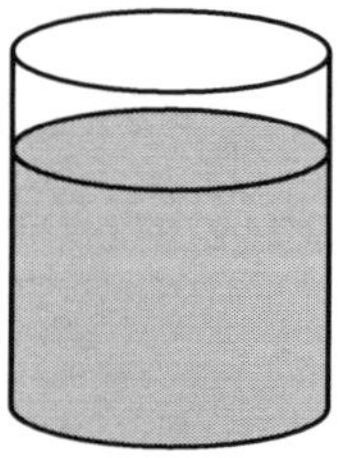

barrel

(2 points)

(b) The barrel contained 20 quarts at first. Jason used 12 quarts of water to wash his car.

How many quarts of water are left? ________ quarts

(2 points)

31. A comb costs $0.65.
Color the number of coins you will need to pay for it.

(2 points)

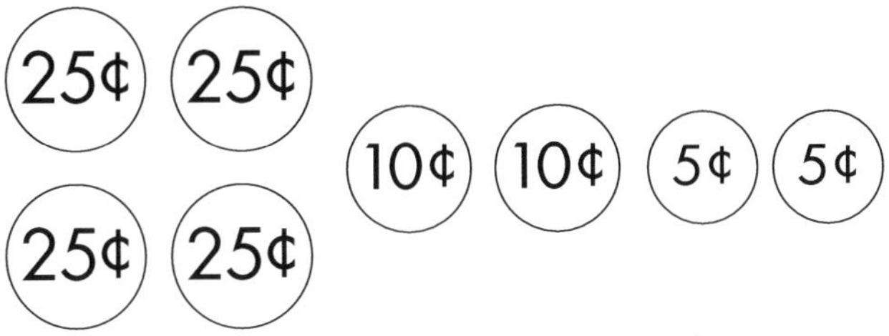

32. Which of the following has the largest value? Circle it.

(2 points)

35 cents

$0.30

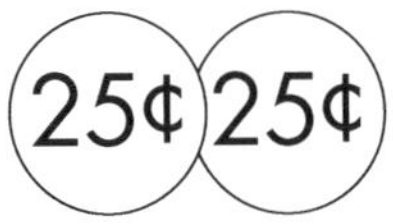

eighteen cents

33.

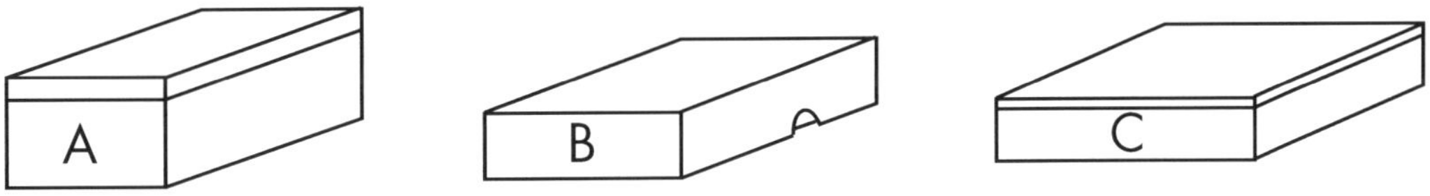

8 kg 3 kg 7 kg

Box C is lighter than box ________.

Altogether the boxes have a weight of ________ kilograms.

(4 points)

34. Study this graph.
Then fill in the blanks. (8 points)

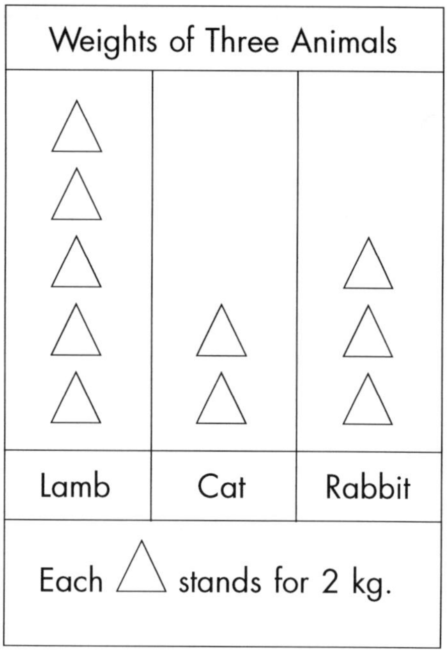

The ________ is the lightest animal.

The weight of the lamb is ________ kilograms.

Section B (20 points)

Do the following problems.

35. Sally's mother made 100 cupcakes.
Sally took 38 cupcakes to school.
How many cupcakes were left at home? (4 points)

________ cupcakes were left at home.

36. Nelly's father gives her $3 a day.
How much will she receive in 1 week? (4 points)

In 1 week she will receive ________.

37. Darla puts 18 ounces of sugar in 1 can and 24 ounces of sugar in another can.
How many ounces of sugar are in the two cans?
(4 points)

There are ________ ounces of sugar in the two cans.

38. Betty has 400 tarts.

Her neighbor bakes 109 fewer tarts than Betty.

(a) How many tarts does her neighbor bake?

(b) Betty bakes 250 more tarts.

How many tarts do they have altogether? (4 points)

(a) Her neighbor bakes ________ tarts.

(b) They have ________ tarts altogether.

39. Ms. London bought 10 boxes of chocolates.

There were 4 bars of chocolate in each box.

(a) How many bars of chocolate did she have?

(b) After giving away 12 bars of chocolate to her neighbors, how many bars of chocolate would she have left? (4 points)

(a) She had ________ bars of chocolate.

(b) She would have ________ bars of chocolate left.

UNITS 1 to 14

Test Yourself 2

POINTS
/100

Section A (80 points)

1. Write the following in numbers. (2 points)

Seven hundred and sixty	
One thousand	

2. Write the following in words. (2 points)

912	
47	

3. Complete the following sentences. (2 points)

(a) 100 more than 900 is ________.

(b) 10 less than 408 is ________.

4. Fill in the missing numbers. (4 points)

(a) 5¢ = $ ________

(b) 300 + ☐ + 7 = 397

Do the following problems.

(c)
$$\begin{array}{r} 732 \\ +\ 89 \\ \hline \end{array}$$

(d)
$$\begin{array}{r} 600 \\ -\ 534 \\ \hline \end{array}$$

5. Join the numbers in order, beginning with the smallest number. (2 points)

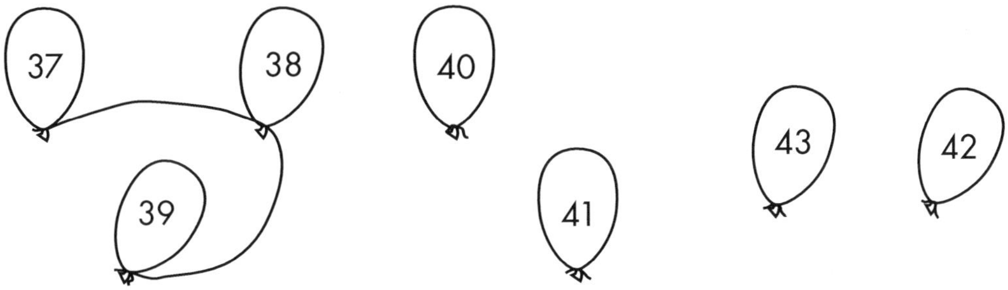

6. Circle the largest number in each set. (2 points)

29	28	40

354	543	435

7. Fill in the missing numbers. (2 points)

76	77	78			81		83	

8. Do the following problems. (2 points)

$$\begin{array}{r} 378 \\ 28 \\ +\ 194 \\ \hline \\ \hline \end{array} \qquad \begin{array}{r} 735 \\ -\ 208 \\ \hline \\ \hline \end{array}$$

9. Three dollars and five cents = $ ________ (1 point)

10. 10 cents = ________ five-cent coins. (1 point)

11. Use a ruler to draw a straight line 3 inches long. (1 point)

12. 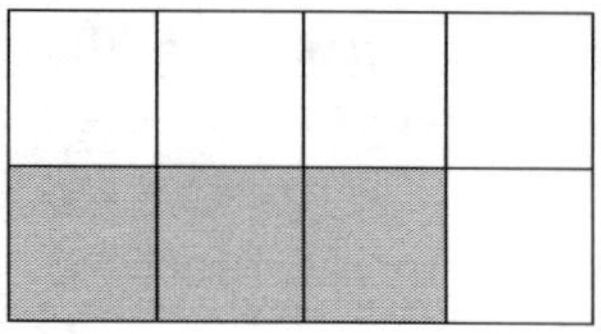

_________ out of _________ equal parts are shaded.

_________ of the figure is shaded.

(3 points)

13. 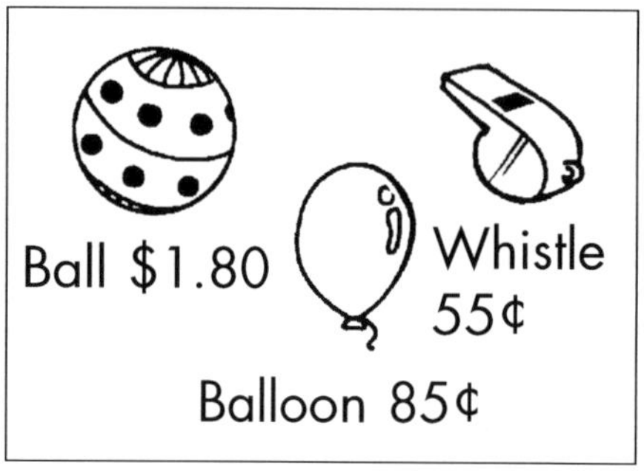

The whistle and the balloon cost _________ altogether.

The ball costs _________ more than the total cost of the whistle and the balloon. (2 points)

14. 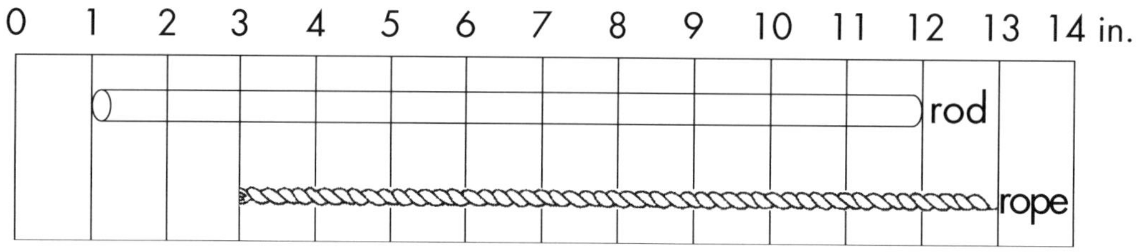

The rod is _________ inches long. (1 point)

The rope is _________ inches long. (1 point)

15. Fill in the blanks. (4 points)

Each  stands for 3 houses.

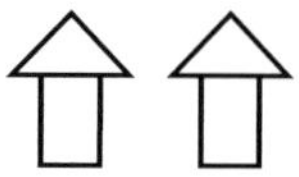 stand for _________ houses.

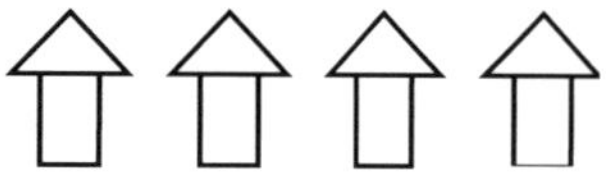 stand for _________ houses.

16. 

What is the time shown on this clock? ________ (1 point)

17. My mom goes to bed at 11:30 p.m.

I go to bed 1 hour before she does.

I go to bed at ________.

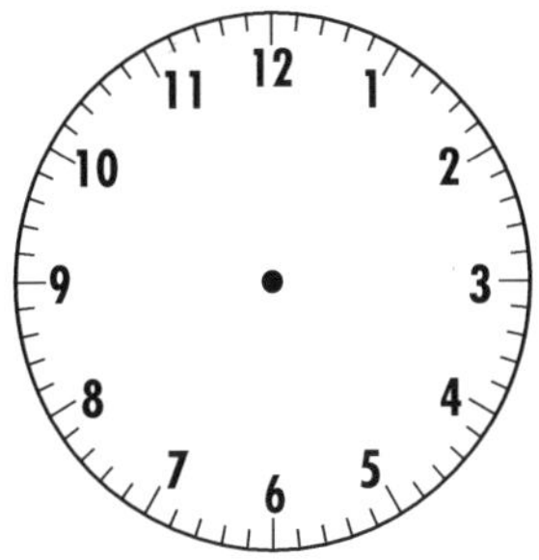

(2 points)

18. Which of the following is equal to 8? Circle it. (1 point)

4 + 2 4 × 2 4 ÷ 2

19. Match the circles to the number sentences. (4 points)

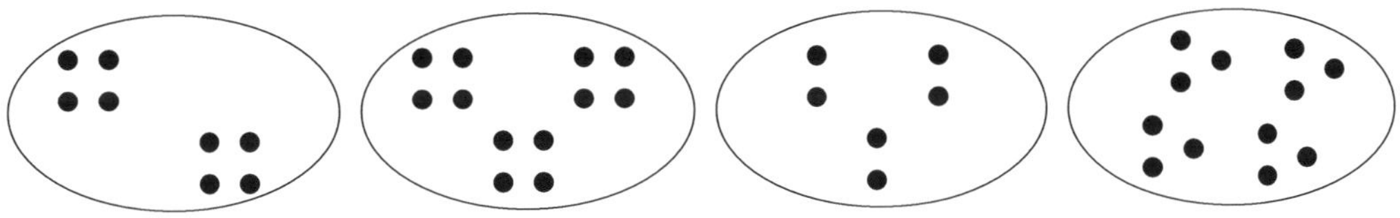

3 × 2 3 × 4 2 × 4 4 × 3

20. Fill in the missing numbers. (2 points)

4 × 4 = ________ 3 × 10 = ________

21.

How many oranges are there? ________ (1 point)

Circle them in sets of 2 oranges.

How many sets are there? ________ (1 point)

22. Write two multiplication sentences for this picture. (6 points)

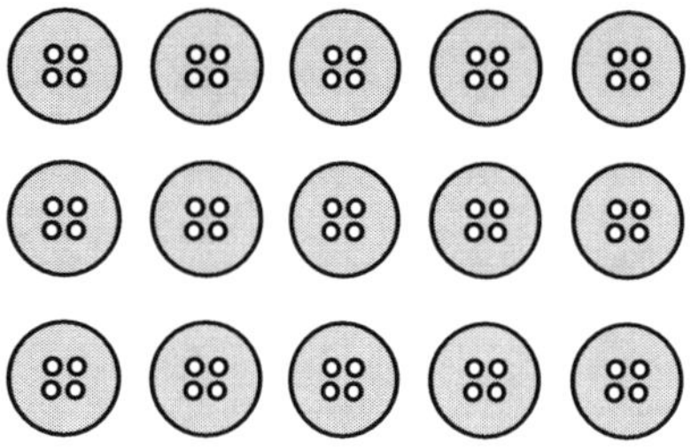

☐ × ☐ = ☐

☐ × ☐ = ☐

23. Write two multiplication sentences and two division sentences about this picture. (12 points)

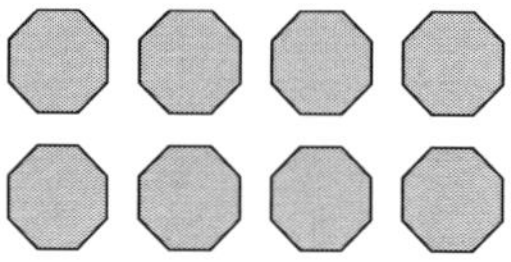

☐ × ☐ = ☐ ☐ × ☐ = ☐

☐ ÷ ☐ = ☐ ☐ ÷ ☐ = ☐

24. Which one of the following figures is **wrong**? Circle it. (1 point)

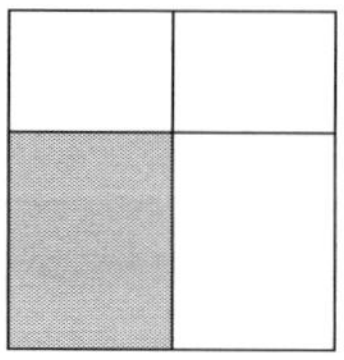

$\frac{1}{4}$

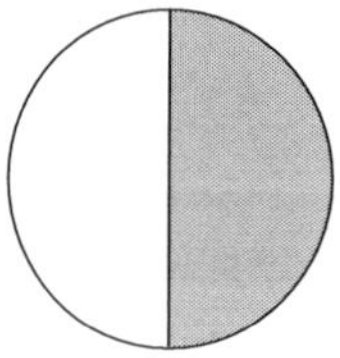

$\frac{1}{2}$

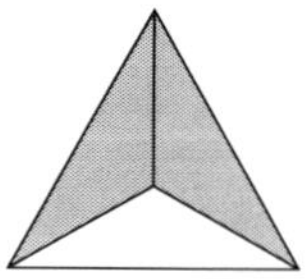

$\frac{2}{3}$

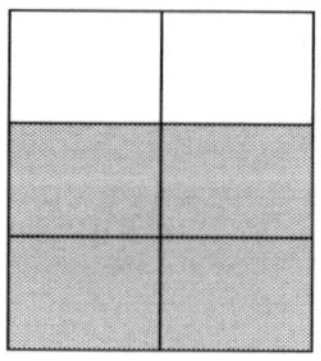

$\frac{4}{6}$

25. Arrange these fractions in order.

Begin with the largest fraction. (1 point)

$\frac{1}{10}, \frac{1}{4}, \frac{1}{6}, \frac{1}{8}$

________, ________, ________, ________.

26. The volume of water in a basin is half the volume of water in the fish bowl.

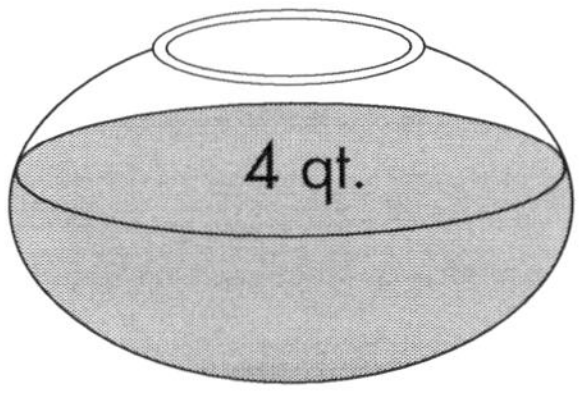

Fish bowl

The volume of water in the basin is

________ quarts. (1 point)

The volume of water in the pail is twice the volume of water in the fish bowl.

The volume of water in the pail is

________ quarts. (1 point)

27. Study these pictures.

Then fill in the blank. (2 points)

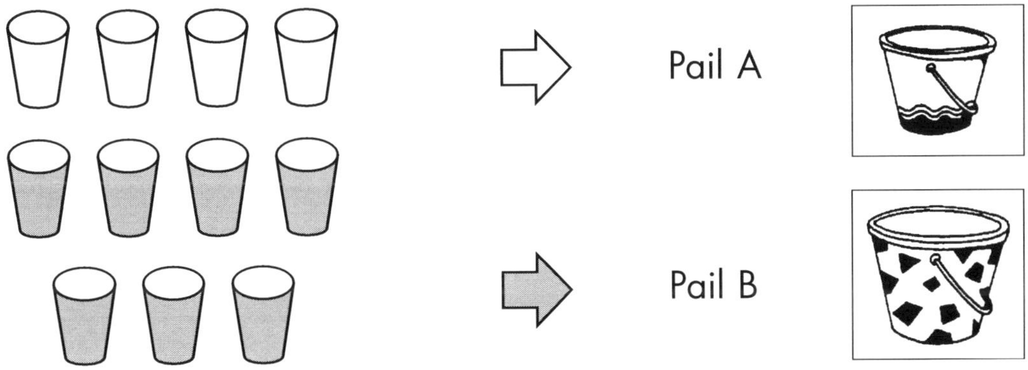

The volume of water in Pail B is ________ glasses of water more than the volume of water in Pail A.

28. Color the next shape in the pattern. (1 point)

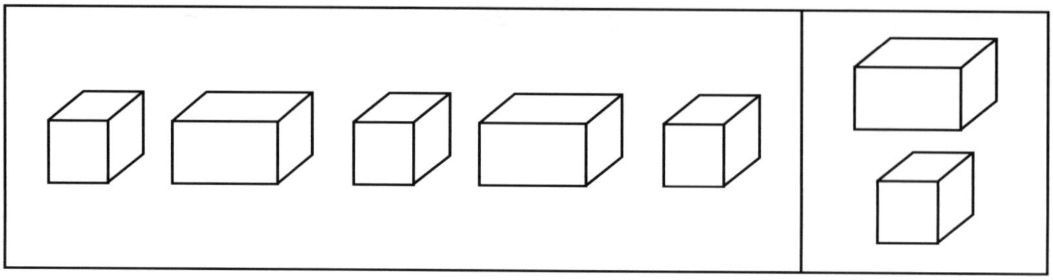

29. Draw the shape that comes next in the box. (1 point)

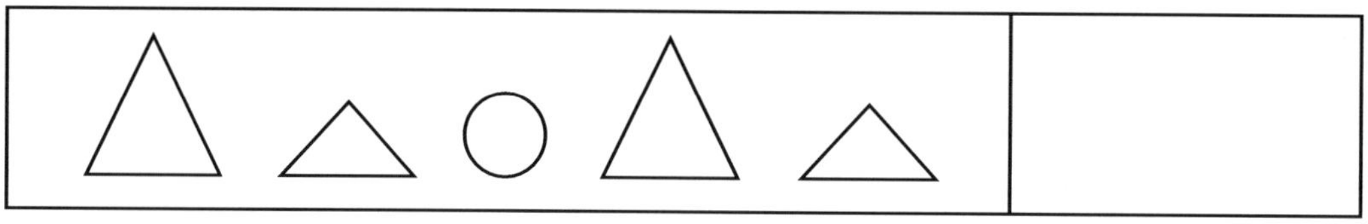

30. 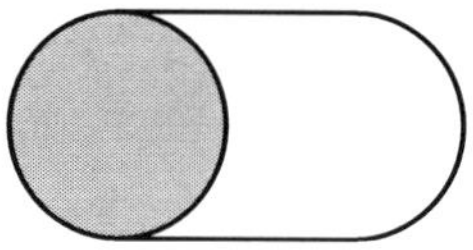 This solid has ________ flat face(s) and ________ curved face(s). (2 points)

31. 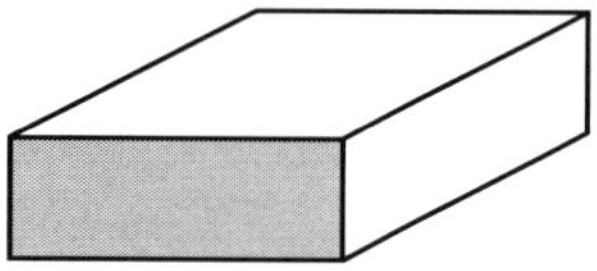 The shape of the shaded face of this solid is a ________. (1 point)

32.  This figure is made up of ________ curves and ________ straight lines. (2 points)

Study this graph.

Then fill in the blanks.

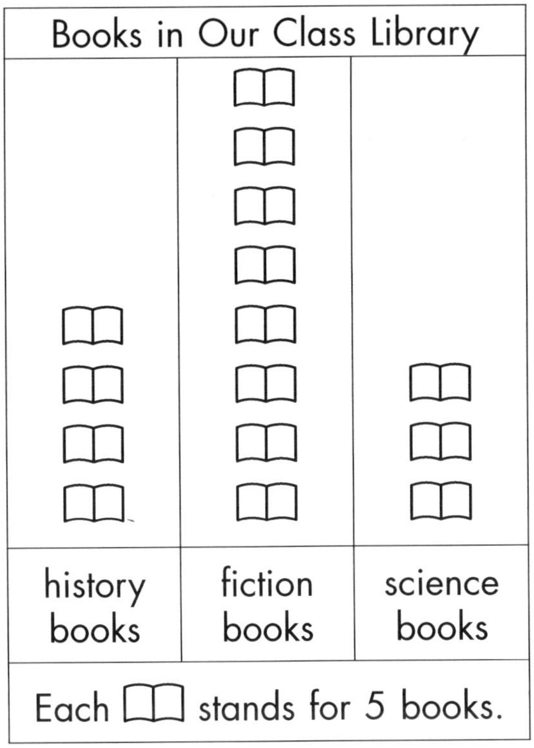

33. There are ________ fiction books. (1 point)

34. There are ________ more history books than science books. (1 point)

35. Altogether there are ________ books in the class library. (1 point)

The graph below shows the amount of money three girls saved. Study it and fill in the blanks.

36. Carol saved $ ________. (1 point)

37. Linda saved $ ________ more than Betty. (1 point)

Section B (20 points)

Do the following problems.

38. Truck A carried 465 gallons of oil.

    Truck B carried 395 gallons of oil.

    How many more gallons of oil did Truck A carry than Truck B? (4 points)

    Truck A carried ________ gallons more than Truck B.

39. Janie puts 40 brownies equally on 4 plates.

    How many brownies are on each plate? (4 points)

    There are ________ brownies on each plate.

40. Kelly had $10.00.

She bought a skirt for $5.00 and a pen for $2.00.

How much money did she have left? (4 points)

She had $ ________ left.

41. Peter has 68 marbles.
He gives his brother 28 marbles and keeps the rest equally in 5 bags.

(a) How many marbles does he keep in 5 bags?

(b) How many marbles are in each bag? (4 points)

(a) He keeps ________ marbles in 5 bags.

(b) There are ________ marbles in each bag.

Study the graph.
Then, fill in the blanks. (4 points)

42.

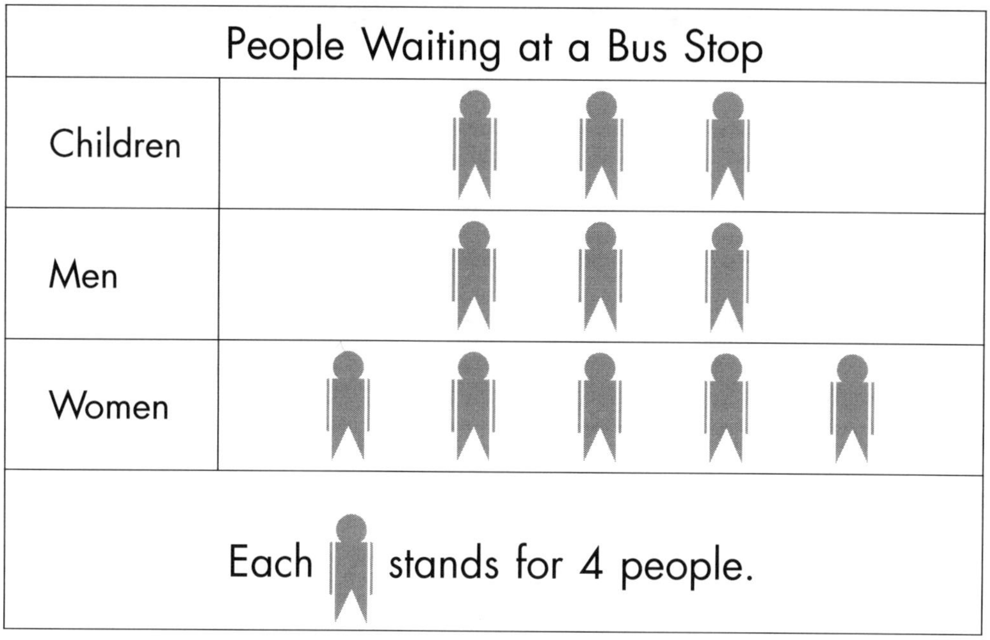

(a) There were as many children as ________ at the bus stop.

(b) ________ women were waiting at the bus stop.

(c) There are ________ fewer men than women at the bus stop.

(d) If 16 people got on the bus, how many people are still waiting at the bus stop?

There are ________ people waiting at the bus stop.

Answers

Unit 1 NUMBERS TO 1,000

1. 248
2. 740 (circled)
3. 290 (circled)
4. (c) ☒
5. 80
6. 200, 600, 1,000, 60
7. 396, 527, 285, 696
8. 653, 1,000, 909, 213
9. hundreds, one, tens
10. 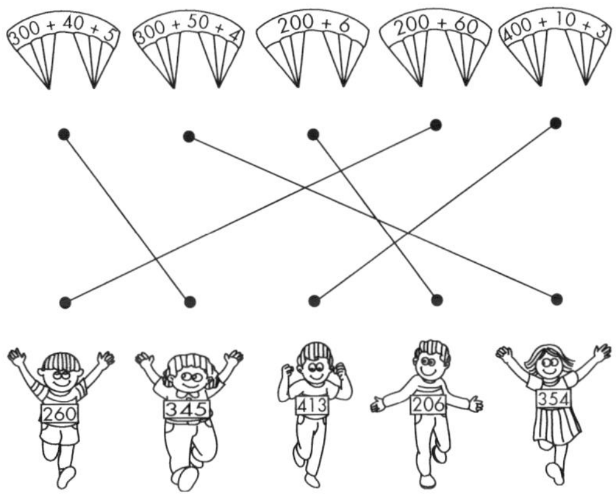

11. 212, 106
12. Eight hundred and thirteen,
    Nine hundred and four
13. 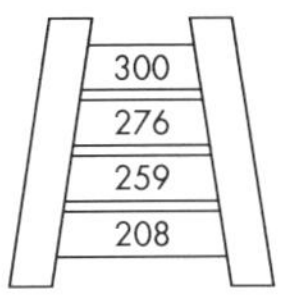

14. 

15. 94, 94, 94
16. 108, 108, 108

Unit 2 ADDITION AND SUBTRACTION WITHIN 1,000

1. 501
2. 420
3. 300
4. 499
5. 657
6. 574
7. 361
8. 436
9. 551
10. 741
11. 346
12. 586
13. 325
14. 257
15. 210
16. 119
17. 230
18. 464
19. 684 – 236 = 448
    He sold 448 apples last weekend.
    448 – 219 = 229
    He sold 229 apples on Sunday.

Unit 3 MULTIPLICATION TABLES OF 2, 3, 4, 5, AND 10

1. 21, 21
2. 20, 20
3. 18, 18
4. 
   6, 6
5. 10, 12, 14, 16, 16, 16
6. 3, 4, 12, 12
7. 4, 5 20, 20
8.  2, 4, 8

    4, 2, 8

   2
9. 6, 9, 12, 15, 18, 21

10. (a) 15, 15 (b) 15, 15

11. (a) 5,

(b) 3, 3, 3, 3, 3, 3,

12. (a) 4, 2, 2, 4 (b) 3, 4, 4, 3

13.

✓	✓	✓	✓	✓	✓	✓	✓	✓
✓	✓	✓	✓	✓	✓	✓	✓	✓

14.

✓	✓	✓
✓	✓	✓
✓	✓	✓
✓	✓	✓
✓	✓	✓
✓	✓	✓
✓	✓	✓

15. 2, 10, 20, 20

16. 14

17. 24

18. 24

19. $3 + 5 = 8$

Ron and Bill bought 8 bags of carrots.

$8 \times 3 = 24$

They bought 24 carrots altogether.

Unit 4 MULTIPLICATION AND DIVISION

1.

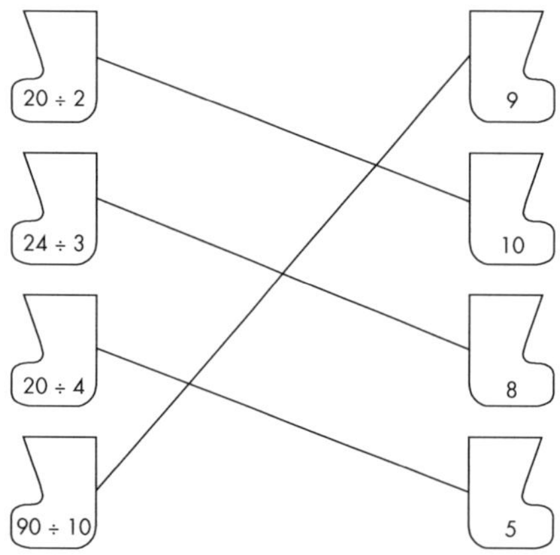

2. (a) 6 (b) 2

3. (a) 20 (b) 30

4. (a) 27 (b) 24

5. 4, 4

6. (a) 4, 3, 12, 3, 4, 12

(b) 12, 3, 4, 12, 4, 3

7. (a) 8 (b) 10

8. (a) 5, 9 (b) 8, 28

9. 15, 3, 5, 5

10. 6, 14, 16

11. 16

12. 40

13. 10

14. (a) 12 (b) 12 (c) 8

15. (a) 32 (b) 4 (c) 4

16. 10

17. 27, 3, 9, 9

18. 3, 4, 12, 12

19. 6

20. 6

21. $6 + 4 = 10$

There were 10 students.

$40 \div 10 = 4$

Each student got 4 pencils.

22. (a) $80 \div 10 = 8$

There are 8 students in each group

(b) $10 \times 3 = 30$

There are 30 girls altogether.

Unit 5 USING MODELS

1. 88, 88
2. 120, 230, 230
3. 98, 65, 33, 33
4. 165
5. 221
6. 420, 128, 280, 828, 828
7. 231
8. (a) 108, 74, 182, 182

   (b) 228, 182, 46, 46
9. 95
10. 157
11. 71
12. 460, 100, 560, 560, 350, +, 460, +, 560, =, 1,370, 1,370

Unit 6 LENGTH

1.

More than 1 m
✓

2.

Less than 1 m
✓

3. 5
4. 3
5. 8
6. 5, 10, 7, B, A, 22
7. (a) B (b) C (c) 2 m (d) 2 m
   (e) B (f) 18
8. (a) 125 (b) 90 (c) 140 (d) 50 (e) 15
   (f) 50 (g) Philip, David, Winnie
   (h) 355
9. 50, 15, 35, 35
10. 400, 250, 650, 650
11. (a) 156 centimeters (b) 160 centimeters
12. 14 meters
13. 22, 5, 17, 17, 22, 17, 39, 39

Unit 7 WEIGHT

1. lighter, heavier
2. watermelon, cantaloupe
3. x, x
4. 3
5. 2
6. 50
7. 30
8. Janet
9. 80
10. 1,600
11. 2
12. 4
13. B, A, 2
14. 900
15. 700
16. 200
17. 500
18. 37, –, 15, =, 22, 22
19. 25 + 12 + 10 = 47
20. Watermelon: 4 lb. Cantaloupe: 4 lb., 1 oz. Honeydew: 3 lb., 15 oz.
21. (a) 360,
    328, 32, 360,
    360
    (b) 688,
    360,
    328, 360, 688,
    688

Unit 8 VOLUME

1. 
2. 
3. B
4. C
5. C
6. B
7. 6
8. (a) 3 (b) 8 (c) 5
   (d) pitcher
9. (a) 10 pails (b) 25 bottles (c) 7 liters
   (d) 3 liters (e) 43 liters
10. 31
11. 54 liters
12. (a) 120 (b) 310
13. 60 liters
14. (a) 188 – 128 = 60

    60 liters were poured into another container.

    (b) 60 – 26 = 34

    Each pail contained 34 liters of water.

Unit 9 MENTAL CALCULATIONS

1. 48, 48, 45, 45
2. 84, 84, 82, 82
3. 76, 76, 71, 71
4. 87, 87, 92, 92
5. 7, 58, 58, 7, 65, 65
6. 43, 43, 45, 45
7. 56, 56, 57, 57
8. 40, 40, 38, 38
9. 63, 63, 59, 59
10. 8, 72, 72, 8, 64, 64

Unit 10 MONEY

1. 

2. 

3. 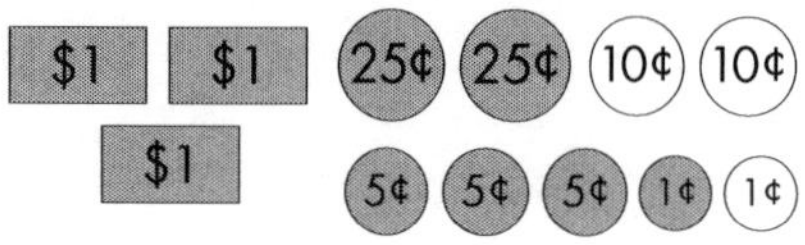

4. (a) (b) (c) (d)

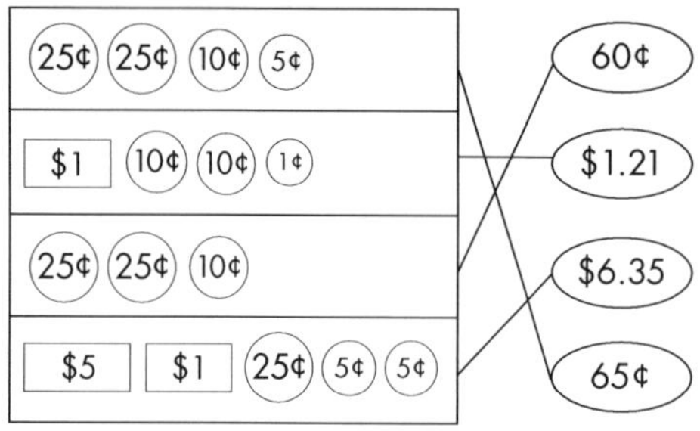

5. (a) 6.30, 305 (b) 2.35, 45
6. (a) 1.25, 13 (b) 2, 0 (c) 40.60

7. (a) (b) (c) (d)

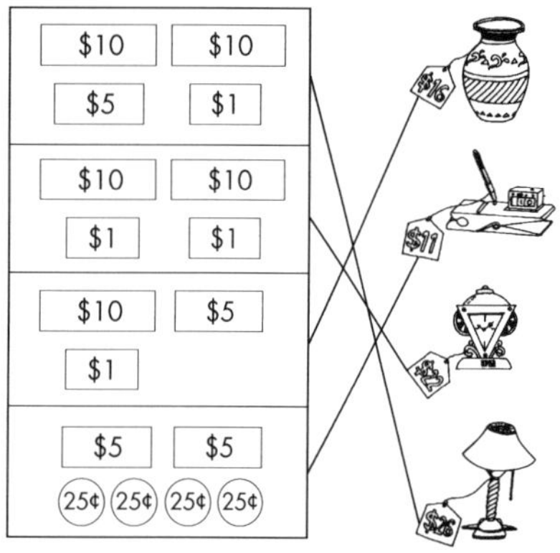

8. $0.70 or 70¢, $27, $11.7, $6.40

9. 15
10. 160
11. 3
12. 55
13. 9
14. 395
15. (a) $3, $7 (b) $4 (c) $2
16. $1
17. $4
18. $5
19. $12 – $3 – $2 = $7

    The teapot costs $7.
20. $50 – $12 = $38

    $38 – $6 = $32

    The handbag cost $32.

Unit 11 TIME

1. 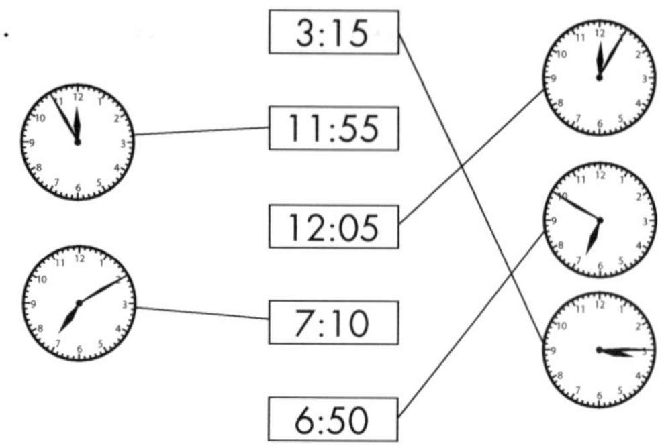

2. 9:20
3. 4:50
4. 7:20
5. 12:30
6. 
7. 
8. 10:15
9. 8:00
10. 3:30
11. 6:15
12. 6:40, 20, 7, 40, 6
13. 8:10, 50, 9, 10, 8
14. 2:35, 25, 3, 35, 2
15. 
16. 
17. 
18. 
19. 9:30 a.m. 
20. 10:40 p.m. 
21. 1:30 p.m. 

Unit 12 FRACTIONS

1. 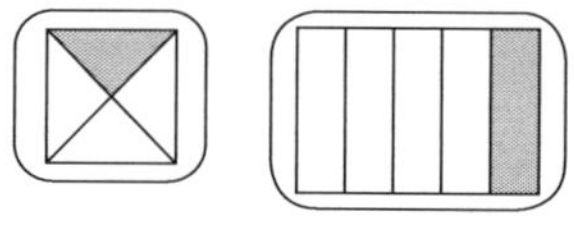

2. $\frac{1}{3}$ $\frac{1}{6}$ 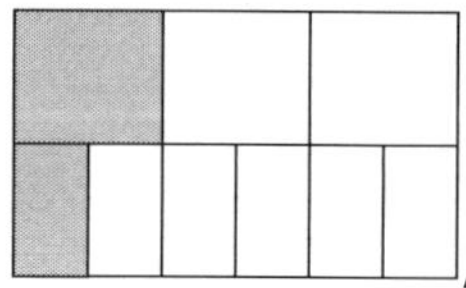 , $\frac{1}{3}$

3. $\frac{3}{4}$ $\frac{5}{8}$ 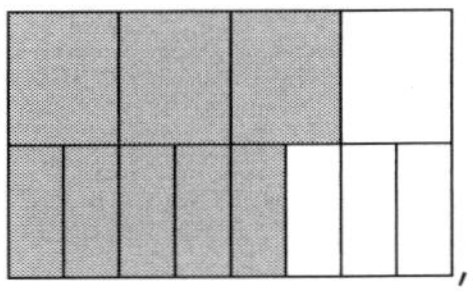 , $\frac{5}{8}$

4. $\frac{1}{2}$ $\frac{1}{6}$ $\frac{1}{4}$ 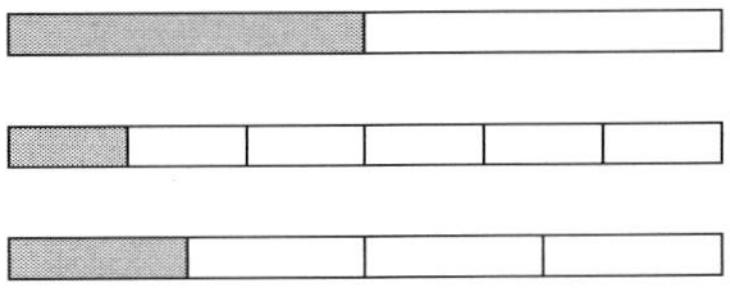 , $\frac{1}{6}$, $\frac{1}{4}$

5. $\frac{3}{8}$ $\frac{5}{6}$ $\frac{7}{12}$ 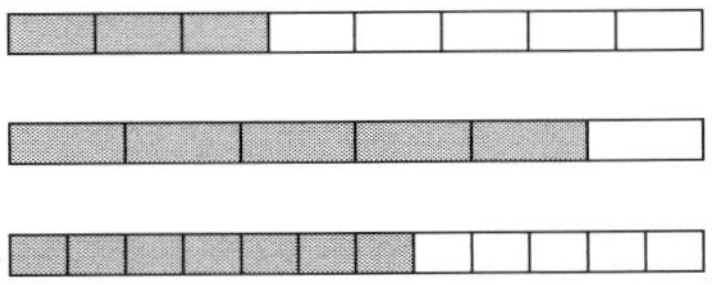 , $\frac{5}{6}$, $\frac{7}{12}$

6. 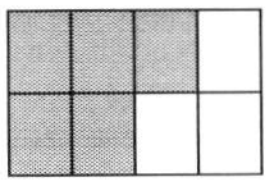

7. 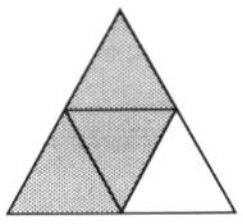

8. 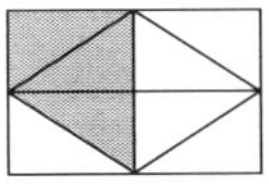

9. 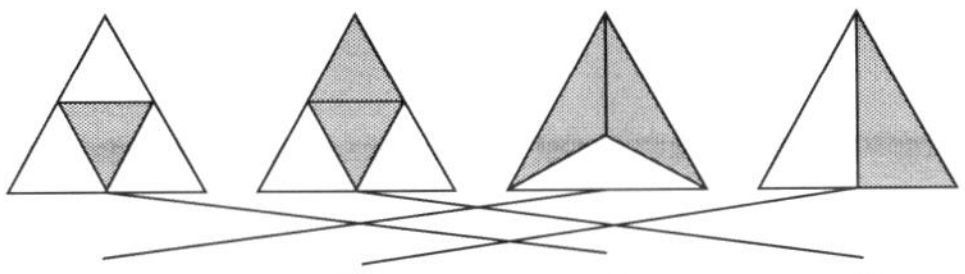

$\frac{2}{3}$ $\frac{1}{2}$ $\frac{1}{4}$ $\frac{2}{4}$

10. 3, $\frac{3}{4}$

11. 5, 8, $\frac{5}{8}$

12. $\frac{3}{5}$

13. $\frac{1}{4}$

14. $\frac{1}{2}$

15. $\frac{5}{6}$

16. $\frac{9}{10}$

17. $\frac{2}{3}$

18. $\frac{3}{4}$

19. $\frac{3}{4}$, $\frac{4}{4}$ or 1 whole, 4

20. 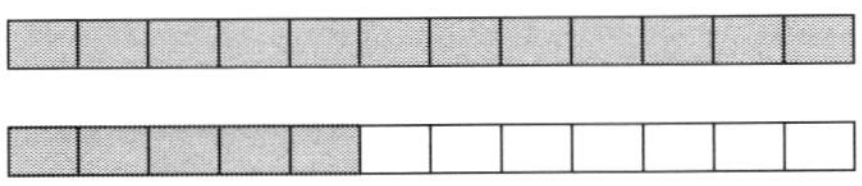 $\frac{7}{12}$

21. ($\frac{2}{5}$) $\frac{2}{4}$ $\frac{3}{5}$

22. $\frac{1}{3}$ $\frac{3}{5}$ ($\frac{3}{8}$)

23. $\frac{1}{3}$ $\frac{1}{4}$ ($\frac{1}{5}$)

24. $\frac{3}{7}$ ($\frac{4}{7}$) $\frac{4}{6}$

25. $\frac{5}{6}$, $\frac{8}{8}$

26. $\frac{1}{2}$, $\frac{1}{3}$

27. $\frac{3}{10}$, 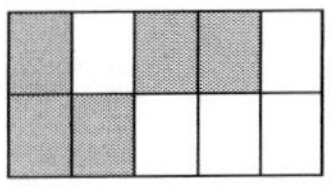

Unit 13 LINES, SURFACES, SHAPES, AND PATTERNS

1. 16
2. 9
3. 
4. 4
5. 13
6. 13, 6, 3, 19
7. 1, 2
8. 2, 1
9. 8, 2
10. 2, 4
11. 1, 2
12. 3, 4
13. flat
14. flat
15. curved
16. 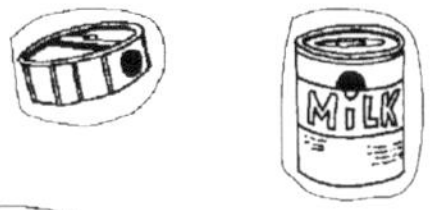

17. 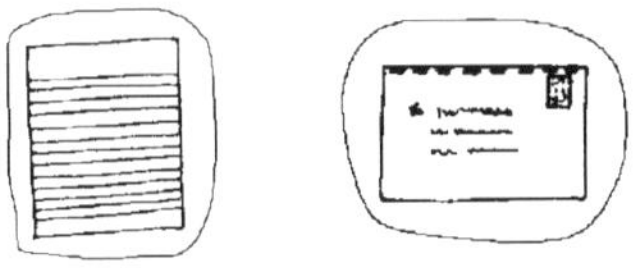
18. circle
19. rectangle
20. triangle
21. square
22. 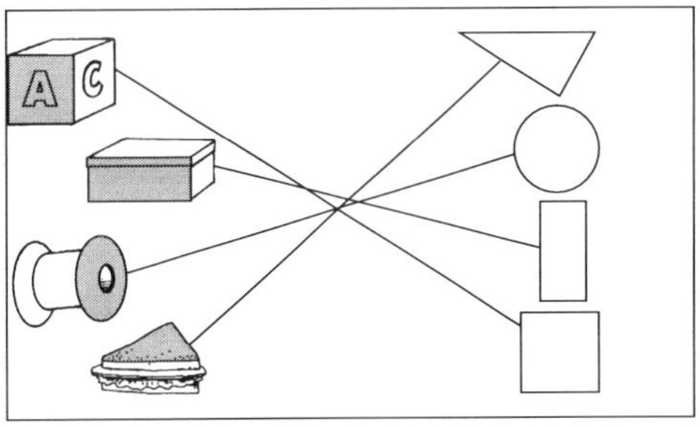

23. triangle, circle, rectangle, square
24. 6, 2
25. (a)  (b) 
(c)  (d) 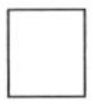
26. 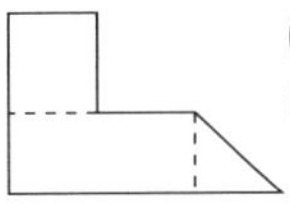 (a) square (b) rectangle (c) triangle

27. 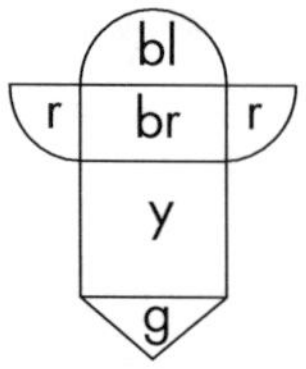

28. 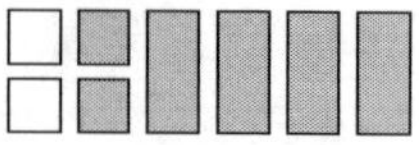

29. 

30. 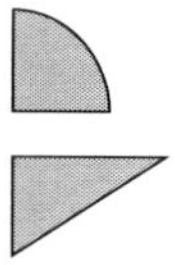

Unit 14 PICTURE GRAPHS

1. (a) 15 (b) 10
2. (a) △△ (b) △△△△
3. 24, 16, 12, 52
4. (a) 6 (b) apple (c) 6 (d) 3
   (e) apple, peach (f) 33
5. 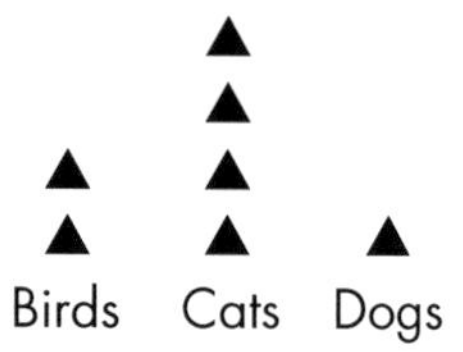

6. (a) 4 (b) 5
   (c) 2 (d) puppy
   (e) duck (f) 3
   (g) 2 (h) hen, puppy
   (i) 3
7. (a) 2A, 2D (b) 4
   (c) 2C (d) 12
   (e)  (f) 10
   2E, 20

Units 1 to 14 TEST YOURSELF 1

1. 35, 100
2. 508, 612, 402
3. (35) 53 51, (129) 200 140
4. 600
5. 7
6. Three hundred and forty-two
7. 40, 50, 60, 90, 100
8. 11, 31, 800, 505
9. 40, –, 21, 19, 40, –, 19, 21
10. 6
11. 121
12. 24
13. 1, 3
14. 6, 3, 18,
    3, 6, 18
15. 12, 20
16. 15, 30
17. 7, 7
18. 8
19. 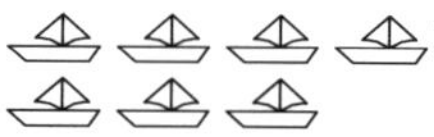
20. 3, 12, 12
21. 7
22. 40, 7
23. 60, 5
24. 36, 10
25. $\frac{11}{12}$
26. $\frac{6}{6}$, $\frac{9}{9}$
27. $\frac{1}{2}$
28. 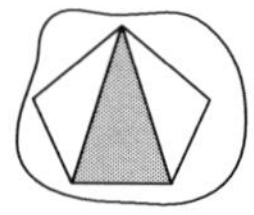
29. 3 a.m., 3 p.m.
30. (a) 4 (b) 8
31. 

32. 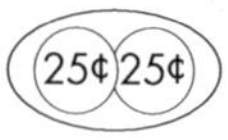

33. A, 18
34. cat, 10
35. 100 – 38 = 62, 62
36. \$3 × 7 = \$21, \$21
37. 18 kg + 24 kg = 42 kg, 42 kg
38. (a) 400 – 109 = 291, 291
    (b) 400 + 250 + 291 = 941, 941
39. (a) 10 × 4 = 40, 40
    (b) 40 – 12 = 28, 28

Units 1 to 14 TEST YOURSELF 2

1. 760, 1,000
2. Nine hundred and twelve, Forty-seven
3. (a) 1,000 (b) 398
4. (a) 0.05 (b) 90 (c) 821 (d) 66
5. 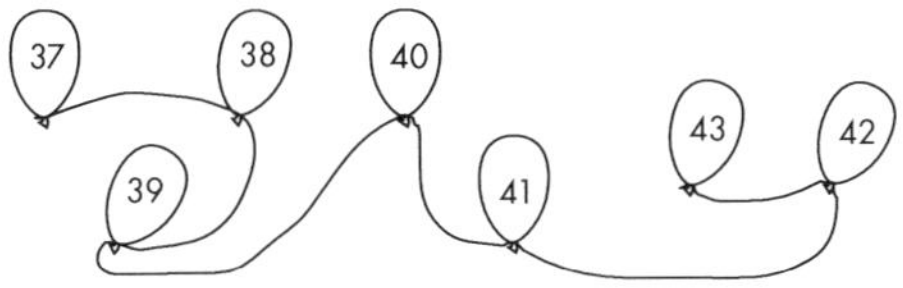

6. (40), (543)
7. 79, 80, 82, 84
8. 600, 527
9. $3.05
10. 2
11. 3 in.
12. 3, 8, $\frac{3}{8}$
13. $1.40, $0.40
14. 11, 10
15. 6, 12
16. 7:55
17. 10:30 p.m. 
18. 4 + 2 (4 × 2) 4 ÷ 2
19. 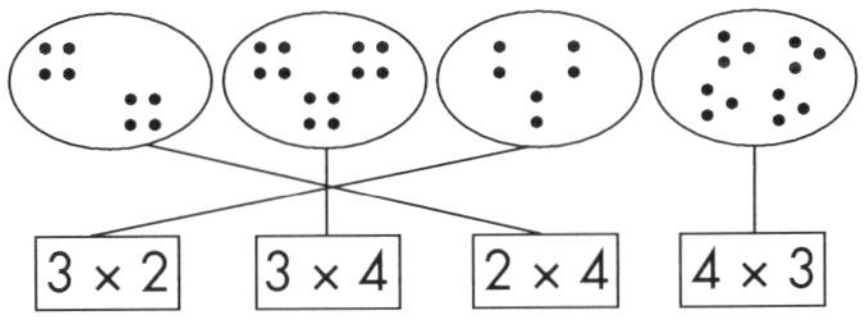

20. 16, 30
21. 6, 3
22. 5, 3, 15
    3, 5, 15
23. 4, 2, 8   2, 4, 8
    8, 2, 4   8, 4, 2
24. 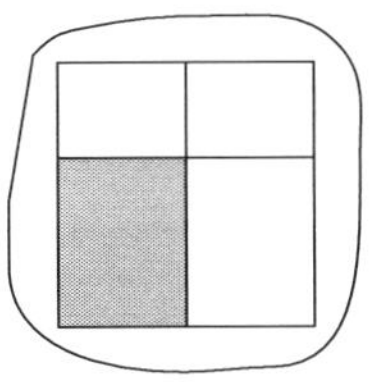
25. $\frac{1}{4}$, $\frac{1}{6}$, $\frac{1}{8}$, $\frac{1}{10}$
26. 2, 8
27. 3
28. 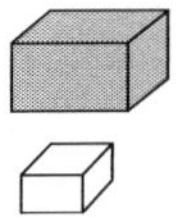
29. 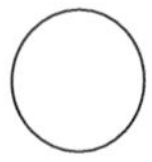
30. 2, 1
31. rectangle
32. 4, 7
33. 40
34. 5
35. 75
36. 15
37. 3
38. 465 – 395 = 70, 70
39. 40 ÷ 4 = 10, 10
40. $10.00 – $5 = $5
    $5 – $2 = $3, 3
41. (a) 68 – 28 = 40, 40
    (b) 40 ÷ 5 = 8, 8
42. (a) men
    (b) 20
    (c) 8
    (d) 3 × 4 = 12

    There were 12 children and 12 men at first.

    12 + 12 + 20 = 44

    There were 44 people at first.

    44 – 16 = 28, 28

Notes

Notes

Notes

Notes

Notes